RADICAL CHEMISTRY

ELLIS HORWOOD SERIES IN ORGANIC CHEMISTRY
Series Editor: Dr J. MELLOR, University of Southampton

Belen'kii, L.I.	**CHEMISTRY OF ORGANOSULFUR COMPOUNDS: General Problems**
Beyer, H. & Walter, W.	**COMPREHENSIVE ORGANIC CHEMISTRY**
Blazej, A. & Kosik, M.	**PHYTOMASS: A Raw Material for Chemistry and Biotechnology**
Brown, S.S. & Kosik, M.	**TOXICOLOGY OF METALS: Clinical and Experimental research**
Cervinka, O.	**ENANTIOSELECTIVE REACTIONS IN ORGANIC CHEMISTRY**
Corey, E.R., Corey, J.Y. & Gasper, P.P.	**SILICON CHEMISTRY**
Dunn, A.D. & Rudorf, W.D.	**CARBON DISULPHIDE IN ORGANIC CHEMISTRY**
Foa, V., Emmet, E.A., Moroni, M. & Colonski, A.	**OCCUPATIONAL AND ENVIRONMENTAL CHEMICAL HAZARDS**
Fresenius, P.	**ORGANIC CHEMICAL NOMENCLATURE**
Hudlicky, M.	**CHEMISTRY OF ORGANIC FLUORINE COMPOUNDS: A Laboratory Manual, 2nd (Revised) Edition**
Hudlicky, M.	**REDUCTIONS IN ORGANIC CHEMISTRY**
Kova, J., Krutosikova, A. & Kada, R.	**CHEMISTRY OF HETEROCYCLIC COMPOUNDS**
Lukevics, E., Pudova, O. & Sturkovich, R.	**MOLECULAR STRUCTURE OF ORGANOSILICON COMPOUNDS**
Lukevics, E. & Zablocka, A.	**NUCLEOTIDE SYNTHESIS: Organosilicon Methods**
Maksic, Z.B.	**MODELLING OF STRUCTURE AND PROPERTIES OF MOLECULES**
Manitto, P.	**BIOSYNTHESIS OF NATURAL PRODUCTS**
Mason, T.M.	**PRACTICAL SONOCHEMISTRY: User's Guide to Applications in Chemistry and Chemical Engineering**
Neidleman, S.L.& Geigert, J.	**BIOHALOGENATION: Prnciples, Basic Roles and Applications**
O'Hagan, D.	**THE POLYKETIDE METABOLITES**
Pizey, J.S.	**THE POLYKETIDE METABOLITES SYNTHETIC REAGENTS, Volume 3**
Sakurai, H.	**ORGANOSILICON AND BIOORGANOSILICON CHEMISTRY**
Smith, E.H.	**COMPARATIVE OXIDATION OF ORGANIC COMPOUNDS**
Smith, K.	**SOLID SUPPORTS AND CATALYSTS IN ORGANIC SYNTHESIS**
Traven, V.F.	**FRONTIER ORBITALS AND PROPERTIES OF ORGANIC MOLECULES**
Trinajstic, N., Knop, J.V., Muller, W.R. & Szymanski, K.	**COMPUTATION CHEMICAL GRAPH THEORY**
Tundo, P.	**CONTINUOUS FLOW METHODS IN ORGANIC SYNTHESIS**
Vernin, G.	**THE CHEMISTRY OF HETEROCYCLIC FLAVOURING AND AROMA COMPOUNDS**

RADICAL CHEMISTRY

M. JOHN PERKINS
Department of Chemistry, Brunel University

ELLIS HORWOOD
NEW YORK LONDON TORONTO SYDNEY TOKYO SINGAPORE

First published 1994 by
Ellis Horwood Limited
Campus 400, Maylands Avenue
Hemel Hempstead
Hertfordshire, HP2 7EZ
A division of
Simon & Schuster International Group

Printed and bound in Great Britain by
Hartnolls, Bodmin.

Library of Congress Cataloging-in-Publication Data

Available from the publisher

British Library Cataloguing in Publication Data

A catalogue record for this book is available from the British Library

ISBN 0-13-320920-2

1 2 3 4 5 98 97 96 95 94

For Pauline

Table of contents

Preface

It was in the mid-1930s that chemists first recognised that reactive free radicals must play a key mechanistic role in some familiar laboratory reactions in organic chemistry. The importance of radical-mediated processes in production of synthetic polymers, especially synthetic rubber, gave a major impetus to the subject during the 1939–45 world war, and although for some time after that the greatest levels of expertise remained in the polymer field, the subject had achieved the status of a mature sub-discipline by the early 1970s. A major two-volume treatise which surveyed free-radical chemistry at that time remains to this day a seminal source-book.[1] Nevertheless, the subsequent two decades have seen significant advances. Particularly important has been the recognition by synthetic chemists that the reputed lack of selectivity so often attributed to free radicals is by no means universal. The result has been the design of complex synthetic strategies depending on radical reactions, many of which have then been reduced to practice with good to excellent yields.

Of at least comparable importance has been the realisation that many biochemical processes are radical in character. In man, for example, reactive radicals frequently mediate in the normal biochemistry of the healthy individual; they also contribute to tissue damage in a catalogue of diseases which seems to be growing in length almost daily.

This short text represents an attempt to provide an overview of the subject suitable both for the advanced student of organic chemistry, and for the qualified practitioner who finds it necessary to brush up on this important facet of the subject. It is based on a series of lectures presented to post-graduate students in England and subsequently in the Philippines during 1992 and 1993; it also owes much to a somewhat shorter course which had earlier been included in an advanced undergraduate programme.

A major part of the challenge to any author of a small specialist work such as this is the presentation of a proper perspective, contemporary to the time of writing, with a sound balance between underlying fundamentals of reaction type and mechanism on the one hand, and modern embellishments, applications, and significance, on the other. Inevitably, there must be an element of subjectivity in

[1] *Free Radicals*, J. K. Kochi, Ed., Wiley, London and New York, 1972.

the selection of examples, and even of topics, and many experts are certain to be disappointed by particular omissions. In this short work on radical chemistry there are indeed many significant omissions, not least the almost complete absence of any discussion of gas-phase systems or of biradicals. The responsibility for all of these – and for what has been chosen for inclusion – is entirely mine.

Chapter 9 constitutes an attempt to feature some topics which have not found a natural home elsewhere in the text. Other themes included in that chapter extend brief introductory remarks which will be found elsewhere in the book: whilst the result may seem somewhat disjointed, it is felt that most readers will discover there something of interest – and if some recognise occasional illustrations, either in Chapter 9 or elsewhere, chosen from my own small contribution to the field, and feel that these do not always provide the very best illustration of the point under discussion, I make no apology!

Proper understanding comes only with practice, and several examples are therefore presented as problems at the end of the book. References are given to the original reports for most of these. Solutions to many of them should be within the compass of anyone who has fully digested the material in Chapters 1–5.

An attempt further to enhance the value of the text has been the inclusion of a bibliography and guide to further reading, principally citing key review articles where readers can seek greater detail, more extensive discussion, or additional examples.

A word about units: most well-known modern basic organic chemistry texts use the kilocalorie as their unit of energy. Papers in many of the world's leading primary journals still follow the same practice, and I have adopted this convention although I know that many will regard this as a retrograde decision. It seems that the calorie-to-joule conversion factor is yet another number which the student of Chemistry must commit to memory. To facilitate this, it appears as a footnote on alternate pages throughout this book. The SI unit of magnetic induction is the tesla. For hyperfine couplings in ESR spectra, the gauss is widely retained, and this practice has been followed in Chapter 5 (N.B. 10 gauss = 1 mT).

In conclusion, I should like to mention my considerable indebtedness to Professor B. Halliwell, and Drs H. Kaur and B. P. Roberts for their critical reading of sections of the manuscript and their invaluable comments, as well as my friends at Brunel University and in Manila for their hospitality during the period in which the manuscript for this book was completed; in particular, I should like to thank Marissa Noel and Titos Quibuyen, as well as the chemistry staff and students of Ateneo de Manila University, for taking so much trouble to ensure that my four months in the Philippines were consistently enjoyable.

M. J. P.
London
February 1994

1

Introduction

Originally, the term "radical" was used to refer to some unchanging part, or "root" of a molecule as the molecule undergoes chemical change. Thus, in the sequence $C_2H_5Br \rightarrow C_2H_5CN \rightarrow C_2H_5CO_2H$, the ethyl "radical" remains unaltered. For a short time in the middle of the nineteenth century it was believed that the ethyl radical, C_2H_5, could be liberated from its compounds as a distinct chemical substance – for instance by reacting C_2H_5Br with certain metals. However, it was soon recognised that the gaseous product of these reactions was not a "free radical" but was in fact a mixture of ethane, ethene, and butane. These and related observations forced the conclusion that carbon was invariably tetravalent in all of its compounds.

Therefore general scepticism greeted Moses Gomberg's demonstration, in 1900, that in oxygen-free solution in benzene the triphenylmethyl radical (**1**) is in equilibrium with its dimer. In fact, although Gomberg's results afforded the first demonstration of trivalency for carbon atoms, they did not provide the first example of what we might now call a stable organic free radical. The species (**2**), porphyrexide, an example of an organic "nitroxide" (see Chapter 5), which was represented in early work as (**2a**), had already been known for more than a decade.

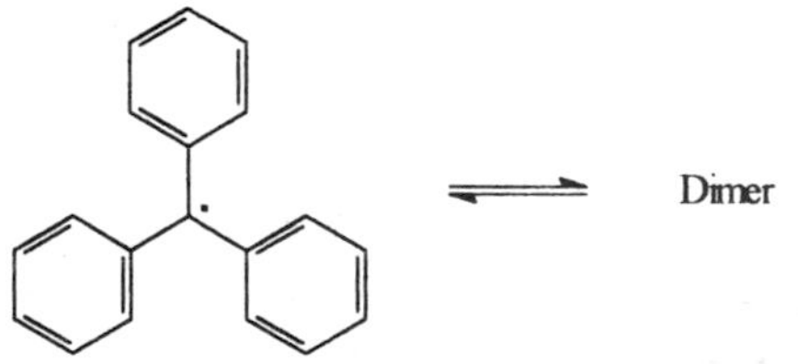

Dimer

(**1**)

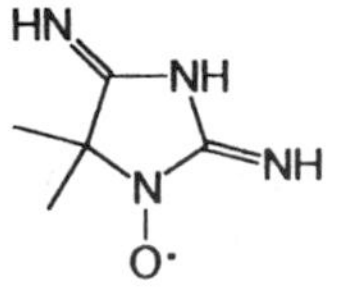

HN NH NH N O·

HN NH NH N O

(**2**) 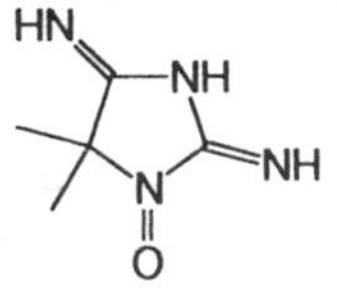 (**2a**)

(1) 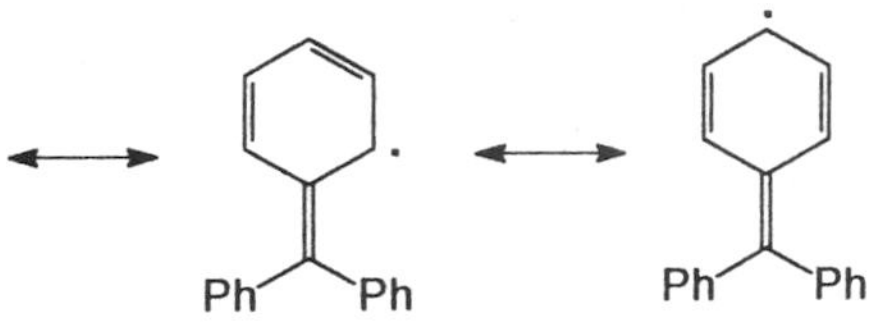 ⟷ (1-*o*) ⟷ (1-*p*)

In modern terms, the triphenymethyl radical is seen as a hybrid of structure (**1**) and resonance (canonical[1]) forms (e.g. **1-*o*** and **1-*p***) in which the unpaired electron is delocalised into the *ortho* and *para* positions of the three benzene rings. The resulting resonance stabilisation is attenuated to some extent by the fact that, in order to avoid steric interference between the phenyl groups, the preferred conformation is such that these groups are twisted out of plane like the blades of a propeller (Fig. **1.1**). Reactivity at the central carbon is therefore reduced, not only by electron delocalisation, but also by steric congestion above and below the molecular plane. This crowding is further increased by any reaction at the central carbon which gives a tetrahedral product. Despite this steric argument, it was not until more than 60 years after Gomberg's original discovery that the dimer of (**1**) was demonstrated to be (**4**), and not hexaphenylethane (**3**).

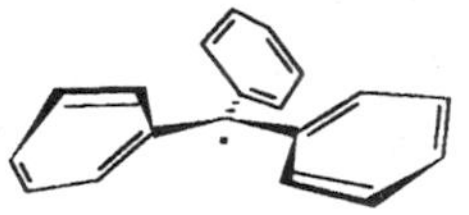

Fig. 1.1

$Ph_3C{-}CPh_3$
(**3**)

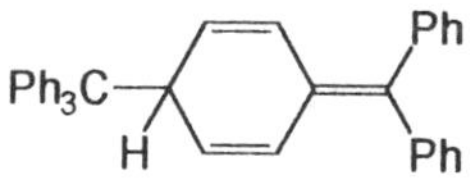

(**4**)

Other stable free radicals were reported from time to time during the thirty years following Gomberg's discovery. In retrospect therefore, it may seem surprising that the possibility that more reactive radicals might have at least some fleeting existence in solution was completely rejected. The breakthrough came almost simultaneously in laboratories on either side of the Atlantic. In Manchester, England, Donald Hey was investigating reactions in which a hydrogen atom of a benzene ring is replaced by a second aromatic grouping, usually phenyl. It became clear that in these "phenylation" reactions, substituents already present in the benzene ring were not

[1]The less frequently used "canonical" means "obeying the rules", i.e. the strict classical rules of valence, in which atoms are connected by formal single, double, or triple (but not partial) bonds.

1 cal = 4.184 J

exerting their usual directing influence, as would have been expected by analogy with other, well-documented aromatic substitution reactions such as nitration.

Ph·
Phenylation

Irrespective of whether the existing substituent was electron-donating or electron-withdrawing, all three isomeric phenylation products were found in roughly comparable quantities (Table 1.1). Hey was forced to the conclusion that the reactive species which attacked the benzene ring must be the electrically neutral phenyl free radical.

Table 1.1: Proportions of isomeric biphenyls obtained in the radical phenylation of simple benzene derivatives.*

	% *o*-	% *m*-	% *p*-
PhOMe	48	32	20
PhMe	52	26	20
PhCl	55	27	18
$PhNO_2$	61	10	29

*These results are representative of data obtained with modern analytical procedures on reactions using a variety of sources of phenyl radicals. In the early work quantitative analysis was difficult, and although from some compounds all three isomers were obtained, it was in fact more usual that only the *ortho*- (most volatile) and *para*- (highest melting) isomers were obtained pure.

By the sort of coincidence which is not uncommon in the history of scientific discovery, little more than 100 miles away, in Durham, Alec Waters was drawing the same conclusion from parallel investigations.

In Chicago, the experiments of Morris Kharasch were of a quite different kind, but the conclusion was the same – that electrically neutral carbon-centred free radicals were implicated in his reactions. Kharasch had been studying the addition of hydrogen bromide across unsymmetrically substituted alkene double bonds. His results were erratic. Sometimes the addition was quite slow, and the expected Markownikoff product was observed. In other experiments, rapid reactions yielded the products of anti-Markownikoff addition. The latter results were eventually attributed to the presence of impurities which were promoting formation of bromine atoms. These add to the alkene to give a β-bromo-alkyl free radical that then removes the hydrogen atom from HBr, thus regenerating a bromine atom (Scheme 1.1).

$$Br\cdot + BrCH_2CH{=}CH_2 \longrightarrow BrCH_2\dot{C}HCH_2Br$$

$$BrCH_2\dot{C}HCH_2Br + HBr \longrightarrow BrCH_2CH_2CH_2Br + Br\cdot$$

Scheme 1.1: Radical addition of HBr to 3-bromopropene.

Over the following fifteen years or so, there was a rapid growth in qualitative understanding of the nature of reactive free radicals, and of the types of reaction in which they participate (see Chapter 2). But particular attention was paid to those radical addition reactions which occurred in a repeating manner to form multiple adducts – reactions better known as alkene polymerisation (or "vinyl polymerisation" – Scheme 1.2). Such reactions were of great utility, and hence very extensively studied. Much *quantitative* data on alkene reactivity towards different radicals was obtained in this work.

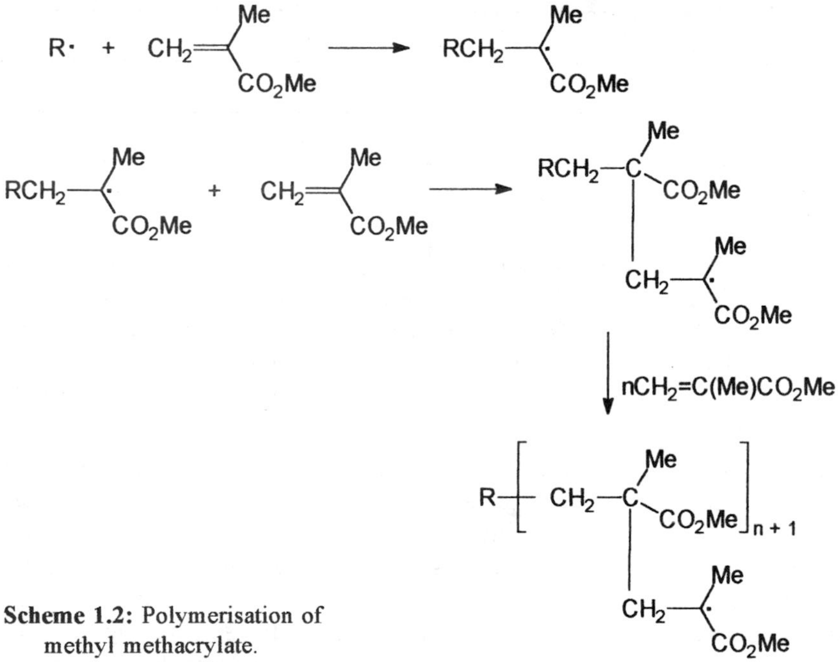

$$R\cdot + CH_2{=}C(Me)CO_2Me \longrightarrow RCH_2\dot{C}(Me)CO_2Me$$

$$RCH_2\dot{C}(Me)CO_2Me + CH_2{=}C(Me)CO_2Me \longrightarrow RCH_2C(Me)(CO_2Me)CH_2\dot{C}(Me)CO_2Me$$

$$\xrightarrow{nCH_2=C(Me)CO_2Me} R[CH_2C(Me)(CO_2Me)]_{n+1}CH_2\dot{C}(Me)CO_2Me$$

Scheme 1.2: Polymerisation of methyl methacrylate.

In contrast, for reactions designed to synthesise *small* molecules, free-radical chemistry seemed to be of little importance. There were a few exceptions, including the well-known Kolbe reaction (Scheme 1.3). In addition, one or two useful reagents were known to react by free-radical mechanisms, for example *N*-bromosuccinimide which has been routinely used to effect allylic bromination [e.g. Equation (1)]. However, the over-riding problem was that reactive free radicals

1 cal = 4.184 J

seemed generally to be uncontrollable. They were *too* reactive. Reference to Table 1.1 is illustrative. All three isomeric products are formed, with very little discrimination. Similarly, whilst students of organic chemistry will very early have learnt something of the utility and mechanism of methane chlorination[1] ($\rightarrow CH_3Cl \rightarrow CH_2Cl_2 \rightarrow CHCl_3 \rightarrow CCl_4$), they are likely also to know that monochlorination of, say, 3-methylheptane will produce significant quantities of *all eight* monochloro-derivatives.

O N–Br O + → O N–H O + Br (1)

$$RCO_2^- \; Na^+ \xrightarrow[\text{(at anode)}]{-e^-} RCO_2\cdot$$

$$RCO_2\cdot \longrightarrow R\cdot + CO_2$$

$$2R\cdot \longrightarrow R{-}R$$

Scheme 1.3: Kolbe electrochemical synthesis of alkanes.

We shall learn in Chapter 4 how this apparent lack of selectivity is far from universal, and in Chapters 6 and 7 how radical chemistry has been tamed to afford realistic, and in some instances preferred, alternatives to ionic reactions with which to accomplish complex transformations. The two examples shown in Equations (2) and (3), without any rationalisation,[2] are presented to whet the reader's appetite by illustrating how a proper grasp of radical chemistry has allowed the subject to develop during the past 10–15 years. The quantitative understanding which facilitated the design of these reactions came from the late 1960s onwards, when fairly reliable kinetic data for a variety of non-polymerisation processes started to become available.

OH HO OBz MeO O Br —SnBu₃, *hν*→ OH HO OBz MeO O (2)

(91%)

[1]The mechanism of chlorination of methane is reviewed in Chapter 2.

[2]See Chapter 6 for a discussion of C-allylation, and a brief explanation of the Mn(III)-promoted reaction.

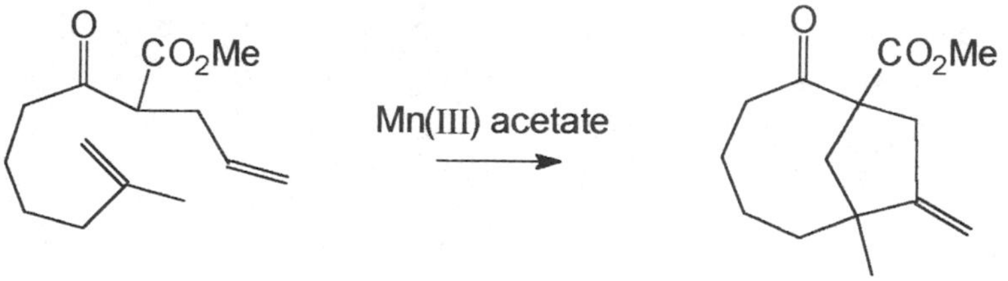

Mn(III) acetate (3)

(70%)

As in almost every field of science, new discoveries have led to modification of old ideas, and the necessity to alter or adapt old terminology. Chemists have tended in recent years to drop the prefix "free" when discussing radical reactions. Nowadays the term "radical" is seldom used in the nineteenth-century sense of the "unchanging molecular root" in a chemical transformation. Instead, a radical has been defined as a molecular entity containing one or more unpaired electrons. In practice, some restrictions are imposed on this all-embracing definition. Thus transition metal ions, or atoms of the alkali metals would seldom be referred to as radicals, although at the other end of the periodic table, halogen atoms are more commonly included. The definition also includes species formed from familiar organic molecules by the loss or gain of a single electron. Examples are the amine *radical cation* (**5**) and naphthalene *radical anion* (**6**). Although radical ions are more generally encountered in the context of oxidations and reductions, they frequently feature as intermediates in processes which also involve neutral radicals. Important examples of this will be encountered in subsequent chapters. The presence of unpaired electrons imparts to radicals the property of paramagnetism; i.e. they are weakly attracted by a magnetic field. In practice, this is observed directly only in cases of stable odd-electron species such as transition metal ions, molecular oxygen, and isolable organic radicals such as porphyrexide.

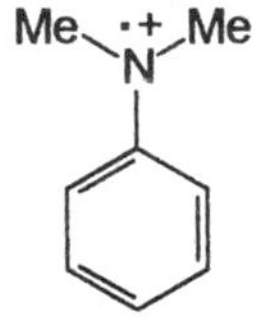

(5) 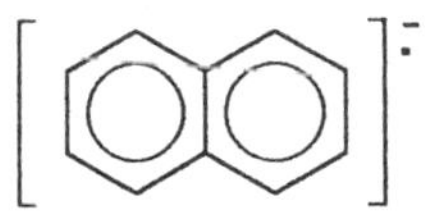 **(6)**

A further development in terminology will be discussed in Chapter 3, where we shall distinguish between "free" and "caged" radicals. The contrast here is between properties of radicals which are diffusing freely and independently through the solution in which they have been generated ("free" radicals), and properties which depend on their immediate molecular environment and, usually, the proximity of a second radical. The importance of this distinction will be developed further in Chapter 5.

1 cal = 4.184 J

This book is designed principally for students of pure chemistry. The significance of radicals in day-to-day life is, however, very great. In the production of many modern materials, notably plastics and other polymers, in the chemistry of paints and foodstuffs, and in the oxidative degradation of lubricants and many other organic materials, as well as in a broad spectrum of biochemical processes in living organisms, radical reactions play a vital role. An understanding of the essential elements of the subject presented in the following chapters should provide the reader with sufficient insight to progress to critical reading of the very extensive literature on these and related topics.

2

Chain reactions and unit steps

2.1 INTRODUCTION

Many chemical transformations involving radicals are "chain reactions". We shall remind the reader of this type of behaviour by reviewing the mechanism of the chlorination of methane mentioned in Chapter 1. Two "unit steps" which are especially important in the production of chloromethane (methyl chloride) are shown in Equations (1) and (2).

$$Cl\cdot + CH_4 \longrightarrow HCl + CH_3\cdot \qquad (1)$$

$$CH_3\cdot + Cl_2 \longrightarrow CH_3Cl + Cl\cdot \qquad (2)$$

In the first step a chlorine atom removes (abstracts) a hydrogen atom from methane to form HCl and a methyl radical. In the second step the methyl radical reacts with molecular chlorine to form chloromethane and a new chlorine atom; evidently the chlorine atom can re-enter the first step and the sequence will then repeat itself. The repetitive character of these steps gives rise to the notion of a "chain" or "radical chain" reaction. The reaction steps represented in Equations (1) and (2) are referred to as **chain-propagating** steps. In both, a radical on the left-hand side is replaced by a radical on the right-hand side.

Whilst the propagating steps may repeat tens, hundreds or even thousands of times, and are the principal product-forming processes, they are insufficient on their own to explain the overall reaction. Chlorine atoms are not formed spontaneously at room temperature; the reaction must be **initiated.** This is most easily accomplished by irradiating the reaction system with light of a wavelength which is absorbed by molecular chlorine and which causes it to dissociate into two chlorine atoms [Equation (3)]. These can then initiate two reaction chains. Continued

$$Cl_2 \xrightarrow{h\nu} 2Cl\cdot \qquad (3)$$

1 cal = 4.184 J

irradiation would generate more and more chlorine atoms and hence more and more radical chains. The overall reaction would become faster and faster until one of the reactants was consumed. In practice the reaction accelerates to a limiting rate, depending on the intensity of the irradiation. This is because of the occurrence of **termination** reactions in which pairs of radicals interact to form non-radical products. In the system described, three termination steps might be expected to compete with the propagation steps. These are shown in Equations (4)–(6).

$$CH_3\cdot \quad + \quad CH_3\cdot \longrightarrow C_2H_6 \qquad (4)$$

$$CH_3\cdot \quad + \quad Cl\cdot \longrightarrow CH_3Cl \qquad (5)$$

$$Cl\cdot \quad + \quad Cl\cdot \longrightarrow Cl_2 \qquad (6)$$

When, under the influence of light, the chlorine molecule dissociates, each chlorine atom carries away one of the two electrons which bonded the atoms together in the molecule. This symmetrical fission of a covalent bond is termed **homolysis**. The alternative dissociative process in which both electrons remain with one partner is **heterolysis**. Reactions which proceed *via* radical intermediates are often referred to as "homolytic reactions". It is worth reminding the reader here that, although heterolysis of hydrogen chloride ($\rightarrow H^+ + Cl^-$) is a process with which we feel completely at ease, in order for it to occur in the gas phase, substantially *more* energy is required (333 kcal mol^{-1}) than is the case for homolytic dissociation ($\rightarrow H\cdot + Cl\cdot$) (103 kcal mol^{-1}). The familiar ionisation reaction is facile only in solution, and then only in solvents such as water, where ionic solvation is so powerful that it more than offsets the heterolytic dissociation energy. Because most radicals are electrically neutral, solvent effects on radical reactions are usually very small. In particular, it should be remembered that water is seldom the problem contaminant that it is for many ionic reactions. As we shall see later, much more serious interference comes from the presence of molecular oxygen.

Of course, the chlorination of methane is complicated by the fact that the initial product, chloromethane, is susceptible to further chlorination to di- tri- and tetra-chloromethanes; the balance between these can be adjusted by the experimental conditions, in particular reactant concentrations. Furthermore, chlorination of methane is a gas-phase reaction. With higher molecular weight hydrocarbons identical chemistry can take place in the liquid phase.

A different type of chain reaction was outlined in Scheme 1.2. In vinyl polymerisation, "chain" is used in two quite distinct senses. Propagation is by a *chain* of addition steps. The result is a long-*chain* molecule or polymer. Initiation of polymerisation (the formation and addition of the R-group in Scheme 1.2) is often by thermal decomposition of an unstable initiator molecule such as a peroxide, rather than by photolysis. Decomposition of the diacyl peroxide, lauryl (bis-dodecanoyl) peroxide, exemplifies this and is shown in Equation (7); the intermediate dodecanoyloxyl radicals are unstable and rapidly decompose ("fragment") into carbon dioxide and undecyl radicals [Equation (8)].

$$C_{11}H_{23}-C(=O)-O-O-C(=O)-C_{11}H_{23} \longrightarrow 2\,C_{11}H_{23}-C(=O)O\cdot \qquad (7)$$

$$C_{11}H_{23}-C(=O)O\cdot \longrightarrow C_{11}H_{23}\cdot + CO_2 \qquad (8)$$

A full discussion of peroxide decomposition and of other initiation processes will be found in the next chapter.

The length of the polymer chain is limited by the occurrence of termination steps – predominantly involving two growing chain radicals interacting to form non-radical products.

In the foregoing discussion we have encountered examples of three major sub-classes of unit-steps in radical chemistry. Two of these sub-classes are bimolecular. These are the termination steps which are **radical–radical** reactions, and the propagation steps which are **radical–molecule** reactions. The third sub-class, represented by the fragmentation of the dodecanoyloxy radical, is **unimolecular**; reaction steps which fall into this third category may also be viewed as chain-propagating, since they are processes in which one radical is transformed into another. We shall now look more closely at these three sub-classes of reaction.

2.2 BIMOLECULAR PROCESSES

(i) Radical–radical reactions

(a) *Radical coupling or dimerisation:* The termination reactions given in Equations (4)–(6) are of this type. When the two interacting species are identical, the term "dimerisation" is correctly applied. "Radical coupling" embraces both dimerisation and the linking of dissimilar radicals [e.g. Equation (5)].

(b) *Disproportionation:* When radicals having a β-hydrogen come together, an alternative to radical coupling exists in which a hydrogen atom is exchanged between the interacting species. This is disproportionation. For example, when lauryl peroxide [see Equation (7)] is allowed to decompose by heating in a relatively unreactive solvent such as benzene, the undecyl radicals undergo competing dimerisation (to give $C_{22}H_{46}$) and disproportionation (to undecane and undecene).

$$2C_{11}H_{23}\cdot \xrightarrow{\text{Dimerisation}} C_{22}H_{46}$$

$$2C_{11}H_{23}\cdot \xrightarrow{\text{Disproportionation}} C_{11}H_{24} + C_9H_{19}CH{=}CH_2$$

1 cal = 4.184 J

Except in the cases of highly resonance-stabilised or otherwise unreactive radicals, the pairing of electrons which occurs when two radicals interact by either of these pathways is a highly exothermic process. Commonly, such reactions are without an activation barrier and occur so rapidly that their rate is determined by diffusive encounter. In mobile solvents such as benzene or water this implies second-order rate constants approaching 10^{10} M^{-1} sec^{-1}.

The factors which determine the competition between two apparently activationless processes (i.e. dimerisation and disproportionation) are still incompletely understood. An important one is rotational diffusion at the point of encounter. For primary alkyl radicals dimerisation usually predominates, but the fraction reacting by disproportionation increases with secondary radicals, and is dominant for tertiary ones.

(ii) Radical–molecule reactions

(a) *Addition:* We have already noted additions to carbon–carbon double bonds that lead to polymers. There are, however, many radical addition reactions in which the products are 1:1 adducts. The anti-Markownikov addition of HBr to alkenes discussed in the 1930s by Kharasch (Chapter 1) constitutes a good example. Another is the addition of CBr_4 to styrene – which normally, on exposure to a radical initiator, undergoes polymerisation. With CBr_4 present, the adduct **(2)** can be obtained in excellent yield (Scheme 2.1). Clearly the rate constant for the transfer of a bromine atom to the initial adduct **(1)** is much greater than the rate constant for addition of **(1)** to a second molecule of styrene.

$$\cdot CBr_3 + PhCH{=}CH_2 \longrightarrow Ph\dot{C}H\text{-}CH_2CBr_3 \;\; \mathbf{(1)}$$

$$\mathbf{(1)} + CBr_4 \xrightarrow{k_{CBr_4}} PhCHBrCH_2CBr_3 \;\; \mathbf{(2)} + \cdot CBr_3$$

$$\mathbf{(1)} + PhCH{=}CH_2 \xrightarrow{k_{styrene}} PhCH(CH_2\dot{C}HPh)CH_2CBr_3$$

Scheme 2.1: $k_{CBr_4} >> k_{Styrene}$

Additions to a wide variety of π-systems have been encountered. However, the success of a projected radical addition depends on the balance between the strength of the π-bond which is broken in the addition, and the strength of the new σ-bond which is formed. Thus additions to the carbon–carbon multiple bonds of alkenes, alkynes and allenes are common, whereas additions to carbonyl groups, which incorporate exceptionally strong π-bonds, are relatively rare.[1]

Additions to conjugated dienes are particularly facile, since the initial adduct is an allylic radical which enjoys a significant measure of resonance stabilisation (Fig. **2.1**). This type of reaction is responsible for the polymerisation of classical paint formulations, such as those based on linseed and other polyunsaturated vegetable oils (Chapter 9), as well as being fundamental to the creation of synthetic rubbers such as neoprene [Equation (9)]. In a similar fashion, addition to styrene generates a radical [e.g. (**1**)] which enjoys benzylic stabilisation.

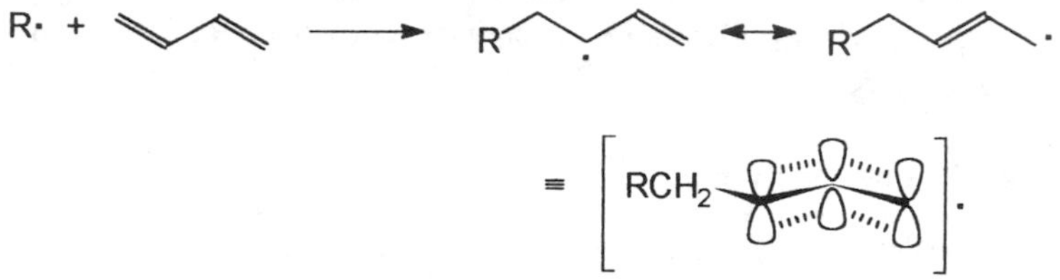

R· + ⟶ R ↔ R

≡ [RCH_2]·

Fig. 2.1: Radical addition to 1,3-butadiene gives a resonance-stabilised allylic radical.

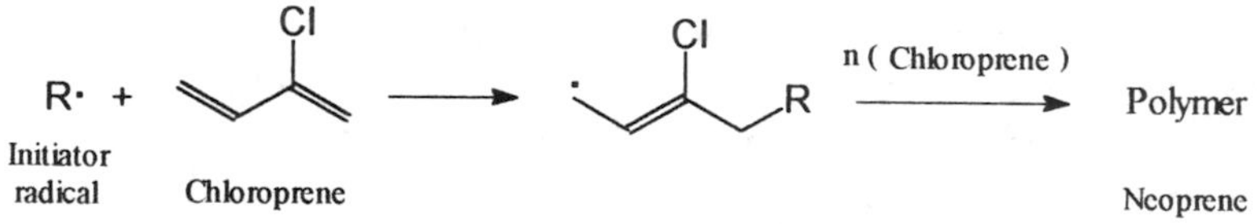

R· + Chloroprene (Cl) ⟶ ·(Cl)R —n (Chloroprene)⟶ Polymer (9)

Initiator radical — Chloroprene — Neoprene

Aromatic π-systems are generally rather unreactive, because addition destroys the aromatic stabilisation. However, it should be recalled that Hey's pioneering experiments involved reactions of this type. We shall see in Chapter 4 that the phenyl radicals (**3**) which mediated in his reactions are particularly reactive. They add to unactivated benzene rings to form substituted cyclohexadienyl radicals, (**4**), akin to the familiar Wheland intermediates of electrophilic aromatic substitution. Oxidation of the cyclohexadienyl intermediate by another radical completes the substitution reaction to give (**5**).

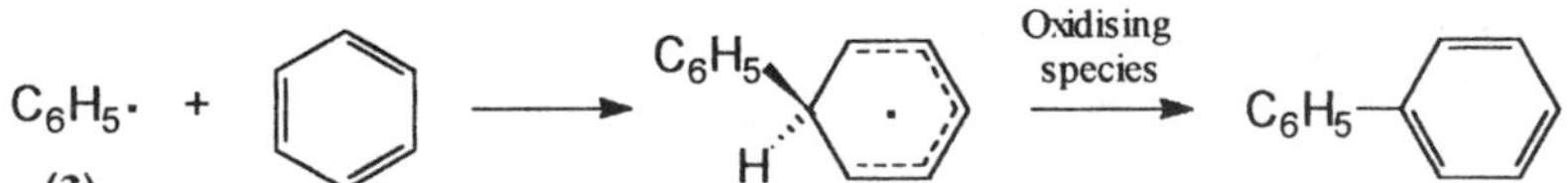

C_6H_5· (**3**) + benzene ⟶ C_6H_5, H (**4**) —Oxidising species⟶ C_6H_5–C_6H_5 (**5**)

[1]An important exception to this generalisation is encountered in *intramolecular* additions. This point is highlighted in Chapter 4, where it is pointed out that intramolecularity can provide a kinetic advantage to a radical addition of a factor of as much as 10^5.

1 cal = 4.184 J

One other addition reaction is of outstanding significance. It is the addition, particularly of carbon-centred radicals, to molecular oxygen. The result is a peroxyl radical ROO·. Oxygen in its ground state is a triplet species, or "biradical"; it contains two unpaired electrons and therefore conforms to our definition of a radical (Chapter 1). Furthermore it behaves like a radical in that the rates of its reactions with alkyl radicals, for example, are close to the diffusion-controlled limit (i.e. $k > 10^9\ M^{-1}\ sec^{-1}$). This is in marked contrast to the rate constants for other radical–molecule additions featured in this section which vary from almost undetectably small to ca. 10^6 or $10^7\ M^{-1}\ sec^{-1}$. (Typical chain-propagating steps in polymerisation, e.g. of styrene, may have rate constants of the order of $10^2\ M^{-1}\ sec^{-1}$.)

(b) *S_H2 reactions, including atom transfer:* Bimolecular reactions in which attack is on a σ-bonded atom are extensively documented. The most familiar examples are the atom-transfer processes, typified by hydrogen-atom abstraction[1] from methane and by chlorine-atom transfer from the chlorine molecule given in Equations (1) and (2). Since these can be designated bimolecular substitution reactions at hydrogen and chlorine respectively, they are symbolised as S_H2 processes, where the subscript stands for homolytic.[2] Atom transfer of halogen from carbon–halogen bonds, particularly those in polyhalomethanes (cf. Scheme 2.1) is also very important. In the case of carbon–iodine bonds the reaction is so rapid that it has been possible to use spectroscopic techniques to examine equilibrium processes such as MeI + Et· ⇌ Me· + EtI.

Hydrogen-atom abstraction from relatively weak C–H bonds by peroxyl radicals (see above), and addition to molecular oxygen of the resulting carbon-centred radicals, comprise the chain-propagating steps of "autoxidation", a process in which an organic compound undergoes oxidation by oxygen. The sequence is represented below by the propagating steps for the autoxidation of cumene. The product is the hydroperoxide (**6**) which has been used commercially as a precursor for phenol.

$$PhC(Me)_2\cdot + O_2 \longrightarrow PhC(Me)_2O{-}O\cdot$$

$$PhC(Me)_2O{-}O\cdot + PhC(Me)_2H \longrightarrow PhC(Me)_2O{-}OH + PhC(Me)_2\cdot$$

$$(\mathbf{6}) \xrightarrow{H^+} [PhOH + Me_2CO]$$

[1]N.B. **Not** "*proton* abstraction"!
[2]S_R2 has also been used (R = radical).

More recently, interest in autoxidation has focused particularly on the reactions of polyunsaturated lipids, because of their importance in biological chemistry and in the food industry (Chapters 9 and 10).

S_H2 reactions are also common at many polyvalent atoms, usually where there is scope for expanding the valence shell, as for example in trialkylboranes or where low-lying *d*-orbitals are accessible. Examples are given in Equations (10)–(12). In some of these cases it has been possible to demonstrate two-step mechanisms by direct detection (using electron spin resonance spectroscopy – Chapter 5) of hypervalent intermediates, for example the phosphoranyl radical, (7), which is on the reaction pathway represented by Equation (12).

$$EtO_2\cdot + Et_3B \longrightarrow Et\text{-}O\text{-}O\text{-}BEt_2 + Et\cdot \qquad (10)$$

$$Me\cdot + PhSeSePh \longrightarrow PhSeMe + PhSe\cdot \qquad (11)$$

$$MeO\cdot + Me_3P \longrightarrow MeOPMe_2 + Me\cdot \qquad (12)$$

$$[MeO\text{—}PMe_3]\cdot$$

(7)

Interestingly, there are very few authenticated examples of S_H2 reactions at carbon. These would be analogous to the familiar bimolecular nucleophilic reactions. Those examples that *are* documented involve cleavage of highly strained small-ring carbocycles, e.g. Equation (13). A mechanism which postulates S_H2 at carbon as a key step in a reaction involving radicals is almost invariably wrong!

Cl· + [bicyclobutane] ⟶ [3-chlorocyclobutyl radical] (13)

Exactly as for addition steps, a very wide range of rate constants has been encountered for S_H2 reactions. The factors which influence the magnitudes of these will be examined more closely in Chapter 4.

(c) *One-electron transfer:* In the presence of suitable oxidising or reducing agents, certain radicals are relatively easily interconverted with the corresponding cations or anions. This is possibly more familiar in the direction of radical formation, particularly by anion oxidation (e.g. the Kolbe reaction; Chapter 1), but

1 cal = 4.184 J

also by cation reduction (e.g. $Ph_3C^+ + V(II) \rightarrow Ph_3C\cdot$). However, there are also many examples in which radicals are oxidised or reduced to cations or anions respectively. The occurrence of such processes has been extensively investigated by electrochemical techniques. Recently it has proved possible to use electrochemical methods to study directly the redox potentials of very low concentrations of some short-lived reactive radicals.

Redox behaviour in many radical processes is frequently mediated by transition-metal ions. Unit steps in these processes may involve one-electron transfer, or they may proceed by ligand transfer, in which a substituent on the metal is directly transferred to the radical. Occasional examples of these processes will be encountered in other chapters; specifically, we shall briefly examine redox behaviour in radical production in Chapter 3 and shall return to the question of ligand transfer and other redox chemistry in Chapter 9.

2.3 UNIMOLECULAR PROCESSES

(a) *Fragmentations:* Decarboxylation of acyloxyl radicals, exemplified in Equation (8), is a radical fragmentation. Acyl radicals, R-CO·, similarly fragment by loss of carbon monoxide, although this process is much slower. It is not observed if R is aryl and is usually negligible if R is a simple primary alkyl group. On the other hand, if R is benzyl, fragmentation to give the resonance-stabilised benzyl radical is rapid. In general, fragmentations are the reverse of additions. When an addition step would be endothermic it is frequently the case that the reverse reaction may be observed. The difficulty of achieving addition to carbonyl groups has been mentioned; the reverse process, fragmentation of alkoxyl radicals, is well known. The fission of a carbon–carbon bond in a t-butoxyl radical to yield acetone and a methyl radical [Equation (14)] is a particularly common example, since numerous peroxides generate t-butoxyl radicals when they decompose (Chapter 3), and this species is also an intermediate in radical-chain halogenation using the reagent t-butyl hypochlorite (t-BuOCl).

$$Me_3C{-}O\cdot \longrightarrow Me\cdot + Me_2CO \qquad (14)$$

When S_H2 reactions pass through detectable intermediates such as (7), the subsequent collapse of these is a fragmentation. Interestingly, in the case of (7) there is an alternative fate to that shown in Equation (12); this also involves fragmentation [Equation (15)]. For related phosphoranyl radicals, the balance between the two fragmentation pathways depends on the nature of the alkyl substituents.

$$\left[MeO{-}P(Me)_3 \right]\cdot \longrightarrow Me\cdot + O{=}P(Me)_3 \qquad (15)$$

(7)

A key step in the interesting rearrangement of the methylenecyclohexadiene (**8**) into (**9**) has been shown to be a radical fragmentation. Scheme 2.2 shows the propagating steps in the radical-chain mechanism which is responsible for this isomerisation; it will be seen that one of these steps is the fragmentation of a substituted cyclohexadienyl radical. This is the reverse of the addition step involved in aromatic phenylation.

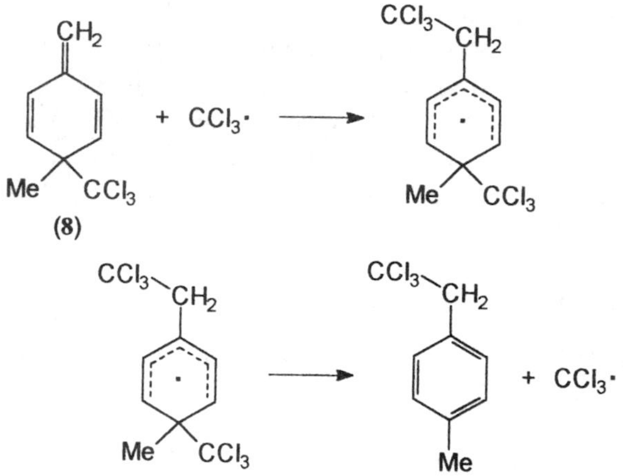

CH_2 + $CCl_3\cdot$ ⟶ CCl_3–CH_2

Me CCl_3 (**8**) Me CCl_3

CCl_3–CH_2 ⟶ CCl_3–CH_2 + $CCl_3\cdot$

Me CCl_3 Me (**9**)

Scheme 2.2

(b) *Rearrangements:* 1,2-Migrations, so common in carbocation chemistry, are relatively rare in radicals. 1,2-Shifts of hydrogen or alkyl have from time to time been claimed, but at the time of writing no authenticated examples are known to the author. For example, neopentyl radicals invariably carry their structural integrity into any products. Groups that *will* undergo 1,2-shifts in radicals have to be able to accommodate the unpaired electron in a low-energy orbital: this is possible where there are accessible vacant *d*-orbitals, as are found with bromine or silicon [Equations (16) and (17)], or where the migrating group has an accessible π-system, as for instance with phenyl or vinyl groups [Equations (18) and (19)]. Cyclopropylcarbinyl radicals, such as the intermediate in Equation (19), if generated directly, will, as expected, rearrange rapidly into the thermodynamically more stable allylcarbinyl isomers. One exception is the cyclopropylbenzyl radical (**10**), which is more stable in the ring-closed form.

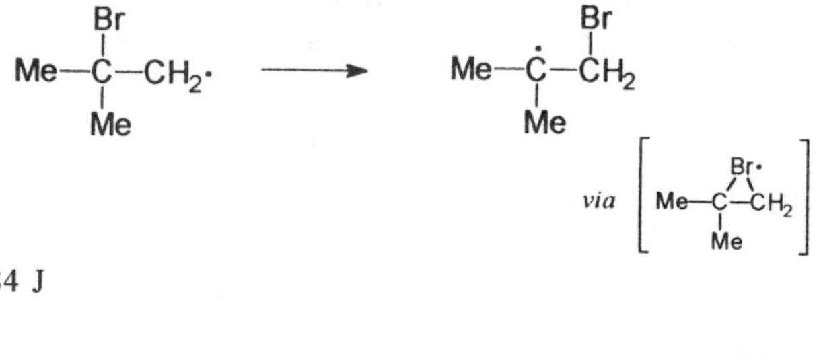

$$\mathrm{Me{-}C(Br)(Me){-}CH_2\cdot \longrightarrow Me{-}\dot{C}(Me){-}CH_2Br} \quad (16)$$

via [Me–C(Br·)(Me)–CH_2 bridged]

1 cal = 4.184 J

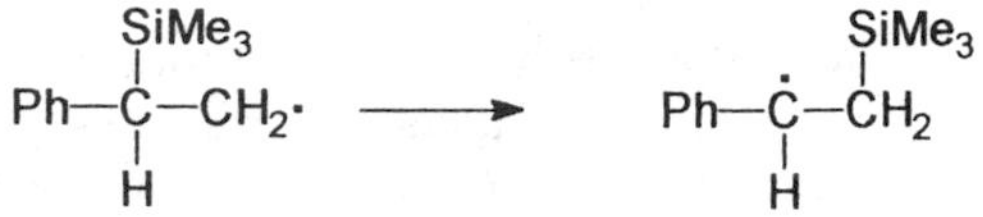

$$\mathrm{Ph{-}\underset{H}{\overset{SiMe_3}{C}}{-}CH_2\cdot \longrightarrow Ph{-}\underset{H}{\dot{C}}{-}CH_2SiMe_3} \qquad (17)$$

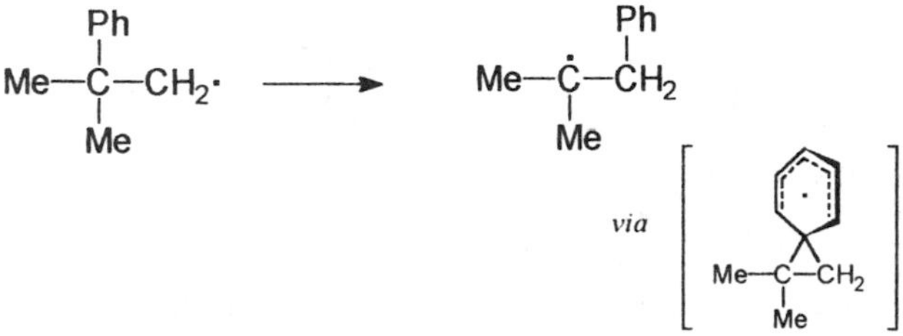

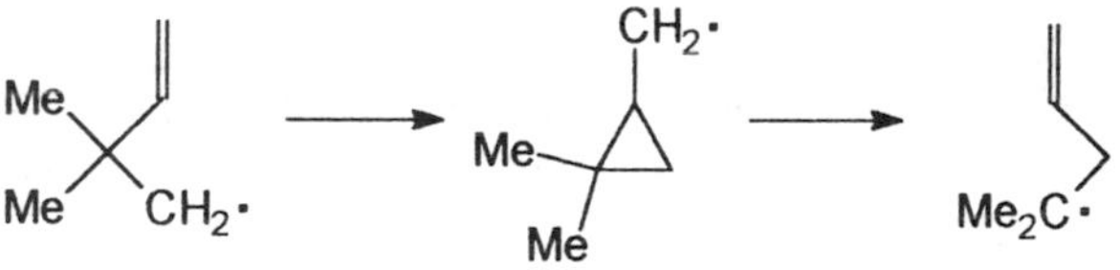

$$\mathrm{Me{-}\underset{Me}{\overset{Ph}{C}}{-}CH_2\cdot \longrightarrow Me{-}\underset{Me}{\dot{C}}{-}CH_2Ph} \qquad (18)$$

via [Me–C(Me)–CH_2 spiro-cyclohexadienyl radical]

$$\mathrm{Me_2C(CH{=}CH_2)CH_2\cdot \longrightarrow Me_2C(cyclo\text{-}C_2H_3CH_2\cdot) \longrightarrow CH_2{=}CHCH_2\dot{C}Me_2} \qquad (19)$$

$PhCH\cdot$–cyclopropyl

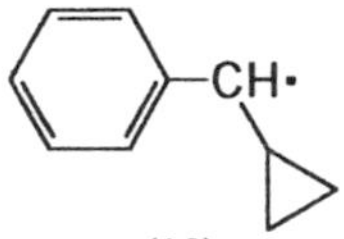

(10)

A 1,2-shift which has attracted considerable attention is that of an acyloxyl group. This reaction is particularly intriguing because the obvious intermediate, **(11)**, which can be generated separately, has been shown *not* to be on the rearrangement pathway (although it does slowly ring-open to the same product radical). On the other hand, isotopic labelling clearly demonstrates that the reaction *does* proceed with transfer of bonding from the ether oxygen to the carbonyl oxygen.

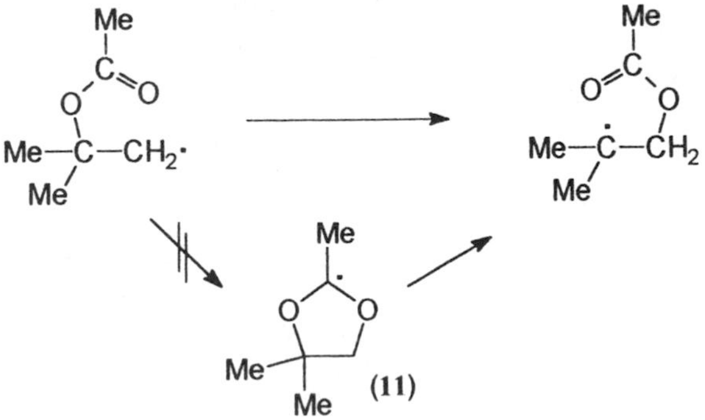

$$\mathrm{Me{-}C(Me)(OC({=}O)Me){-}CH_2\cdot \longrightarrow Me{-}\dot{C}(Me){-}CH_2OC({=}O)Me}$$

(not via) 2-Me-4,4-Me₂-1,3-dioxolan-2-yl radical **(11)** (which does ring-open to the product)

The example of acyloxyl-group migration has been chosen to introduce the "fish-hook" arrows that are commonly used in writing radical mechanisms (Fig. **2.2**). In Fig. **2.2a** all the electron shifts are represented. A frequently encountered alternative representation depicts movement in only one direction (Fig. **2.2b**).

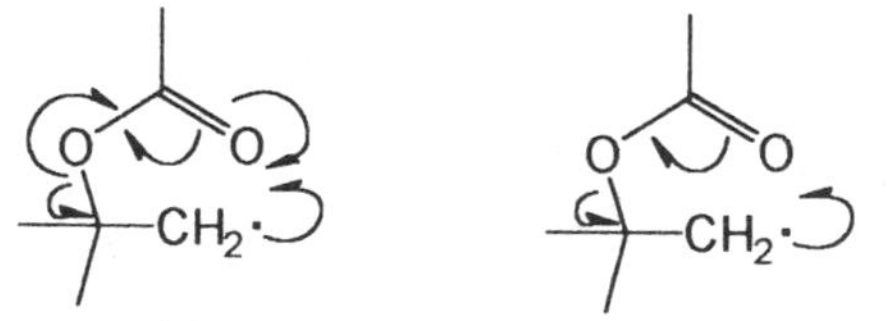

(a) **(b)**

Fig. 2.2: 1,2-Migration of acetoxyl group, using "fish-hooks" to represent one-electron shifts.

1,2-Shifts of vinyl or of phenyl involve initial intramolecular addition of a radical to a neighbouring π-system. Many so-called radical rearrangements stop after this first intramolecular addition step. Thus the cyclisation of the 5-hexenyl radical (**12**) to cyclopentylmethyl, which features prominently elsewhere in this book, whilst often classified as a radical rearrangement, may equally be viewed as an intramolecular example of a radical addition.

(12)

The above cyclisation is characteristic of hexenyl *radicals* and its occurrence in suitably designed systems has frequently been cited as diagnostic of radical intermediates in a reaction mechanism. Other intramolecular radical additions are documented (for examples, see Chapter 6), but the most favourable are those which lead to five-membered or six-membered rings.

Intramolecular atom transfers, e.g. Equation (20), are also classified as rearrangements. Here, the most frequently encountered examples involve six-membered ring transition states.

(20)

1 cal = 4.184 J

2.4 AN INTRODUCTION TO THE KINETICS OF RADICAL CHAIN REACTIONS

Let us examine a simple example – beginning with the decomposition of lauryl (bis-dodecanoyl) peroxide: $(C_{11}H_{23}COO)_2$ (= "**P**"). In boiling benzene, this decomposes [cf. Equations (7) and (8)] with a half-life of ca. 4 h; i.e., in the scheme below, k_1 = ca. 10^{-4} sec^{-1}. To a first approximation, let us assume that the decomposition occurs cleanly, giving two undecyl radicals, and that these do not react with the solvent but dimerise or disproportionate to give $C_{22}H_{46}$ and $C_{11}H_{24}$ plus $C_{11}H_{22}$:

$$\mathbf{P} \xrightarrow{k_1} 2C_{11}H_{23}\cdot \quad (\textit{via}\ 2C_{11}H_{23}COO\cdot)$$

$$2C_{11}H_{23}\cdot \xrightarrow{2k_2} \text{products}$$

Since the second step is diffusion controlled, $2k_2$ is[1] approximately 10^{10} M^{-1} sec^{-1}. The stationary state approximation gives:

$$d[C_{11}H_{23}\cdot]/dt = 2k_1[\mathbf{P}] - 2k_2[C_{11}H_{23}\cdot]^2 = 0$$

whence

$$[C_{11}H_{23}\cdot] = (2k_1[\mathbf{P}]/2k_2)^{1/2}$$

Suppose $[\mathbf{P}]_0$ = 0.005 M, then shortly after the reaction commences:

$$[C_{11}H_{23}\cdot] = (2 \times 10^{-4} \times 0.005/10^{10})^{1/2} = 10^{-8}\ \text{M}$$

Now let us add a *Substrate* which is known to react very rapidly with $C_{11}H_{23}\cdot$ to generate a new radical R·, and suppose that k_p = 10^4 M^{-1} sec^{-1} for the chain-propagating step (i), in which R· is regenerated:

$$R\cdot + \textit{Substrate} \xrightarrow{k_p} \textit{Product} + R\cdot \qquad \text{(i)}$$

The termination is now:

$$R\cdot + R\cdot \xrightarrow{2k_t} \text{product} \qquad \text{(ii)}$$

[1]In general, k is the rate constant for *product formation*. The rate constant for disappearance of radicals in dimerisation or disproportionation is conventionally expressed as $2k$ since two identical radicals are removed.

Taking [R·] to be ca. 10^{-8} M [because $2k_t$ for reaction (ii) is again ca. 10^{10} M^{-1} sec^{-1}] and choosing $[Substrate]_0$ = 1 M, then the initial rate of disappearance of R· in (i) is $10^{-8} \times 1 \times 10^4 = 10^{-4}$ M sec^{-1}, and the rate of disappearance of R· in (ii) is $10^{-8} \times 10^{-8} \times 10^{10} = 10^{-6}$ M sec^{-1}.

Comparing these figures, it is $10^{-4}/10^{-6}$ (= 100) times more likely that R· will disappear in the propagation step (i) than in the termination step (ii) – i.e. the **kinetic chain length** is 100. In other words, for every undecyl radical formed, 100 *Substrate* molecules are transformed into *Product*.

The purpose of this simplified analysis is to show how the diffusion-controlled termination step can be less productive than a propagation step which has a second-order rate constant some six orders of magnitude lower! The value of 10^{-8} M for the stationary-state radical concentration is reasonably representative of what might be expected in a laboratory experiment involving radical intermediates. It is a number to which we shall return in Chapter 5.

Some representative rate constants for a variety of radical reaction steps have been collected in an Appendix at the end of the book.

There is, in conclusion, one important practical point to be remembered – that radical–molecule reactions in which the molecule is *oxygen* have rates close to the diffusion-controlled limit. Only rigorous exclusion of oxygen will prevent it from interfering with other radical processes; in this sense it may be an even less desirable contaminant in many radical-mediated reactions than is water in ionic ones!

1 cal = 4.184 J

3

Initiation and some non-chain processes: "free" and "caged" radicals

Although the necessity for initiation was properly acknowledged in Chapter 2, the emphasis there was very firmly on the importance and variety of propagation and termination processes which are the product-forming steps in radical reactions. The object of the present chapter is to restore the balance by looking much more closely at the range of procedures which are available for initiation. The examples already encountered depend on thermolysis and photolysis of weak covalent bonds. These will be reviewed in some detail, and the importance of ionising radiation and of one-electron transfer will be introduced.

3.1 THERMOLYSIS

Di-t-butyl peroxide (**1**) is a particularly good example of a peroxide initiator because its decomposition is cleanly first-order under most experimental conditions and involves the simple homolysis of the oxygen–oxygen bond. The rate is essentially independent of whether the reaction is studied in solution or in the gas phase. It is generally assumed that the activation energy is in effect the dissociation energy; in other words, a study of the temperature-dependence of the decomposition rate will yield a measure of the O–O bond dissociation energy. The result is ca. 38 kcal mol^{-1}; at 140°C the half-life is ca. 2 h, which actually makes the compound just a little bit *too* stable for many practical applications.

$$\underset{(\mathbf{1})}{\mathrm{Me_3C{-}O{-}O{-}CMe_3}} \xrightarrow{140^{\circ}\mathrm{C}} 2\ \mathrm{Me_3C{-}O\cdot}$$

An alternative source of t-butoxyl radicals is the azo-compound, t-butyl hyponitrite (**2**). Like most other azo-compounds, this undergoes thermolysis by simultaneous rupture of the two bonds on either side of the –N=N– linkage, a reaction which is to a large extent driven by the enormous thermodynamic stability

$$Me_3C{-}O{-}N{=}N{-}O{-}CMe_3 \ \textbf{(2)} \xrightarrow{55^{\circ}C} 2\ Me_3C{-}O^{\bullet} + N_2$$

$$NC(Me)_2C{-}N{=}N{-}C(Me)_2CN \ \textbf{(3)} \xrightarrow{90^{\circ}C} 2\ NC{-}C^{\bullet}(Me)_2 \ \textbf{(4)} + N_2$$

of dinitrogen, N_2. Compound **(2)** decomposes with a half-life of ca. 2 h at 55°C. More familiar azo-compound initiators, like azobisisobutyronitrile ["AIBN"; **(3)**], generate carbon-centred radicals [in the case of **(3)** these are 2-cyano-2-propyl radicals **(4)**], and have decomposition rates (Table 3.1) which depend on the strengths of the bonds being broken. Inspection of Table 3.1 reveals an evident inverse correlation between azo-compound stability and the possibilities for resonance stabilisation in the radicals which are formed on decomposition.

Table 3.1: Relative rates of unimolecular decomposition (80°C) of symmetrical azo-compounds having the general formula: $X(Me)_2C{-}N{=}N{-}C(Me)_2X$

X–	*k*(rel.)
Me–	1.00
$PhCH_2$–	9
PhO–	29
N≡C– **(3)***	7×10^5
EtO_2C–	2×10^7
Ph–	1×10^7

*Half-life at 80°C = ca. 1.5 h.

1 cal = 4.184 J

Table 3.2: Approximate relative rates (at 80°C) of unimolecular decomposition of a representative selection of organic peroxides

Peroxide	k(rel.)
Bu^tOOBu^t (**1**)	0.2
$(CH_3CO_2)_2$ (**8**)	500
$(PhCO_2)_2$ (**9**)	200
$MeCO_2OBu^t$ (**6**)	(1.00)*
$PhCO_2OBu^t$	5
$PhCH_2CO_2OBu^t$ (**5**)	1200
p-$MeOC_6H_4CH_2CO_2OBu^t$	120 000
p-$O_2NC_6H_4CH_2CO_2OBu^t$	180
$Bu^tOO_2CCO_2OBu^t$ (**7**)	10^6
$MeCMe_2CO_2OBu^t$	3500
$EtO_2CCMe_2CO_2OBu^t$	300
$PhOCMe_2CO_2OBu^t$	5×10^7

*Approximate half-life at 80°C = 600 h.

The synchronous fission of two bonds when azo-compounds decompose is mirrored in the decomposition of some peroxides. Rates of decomposition of representative peroxides are indicated in Table 3.2. The much higher rate for t-butyl α-phenylperoxyacetate (**5**) than for the parent peroxyacetate (**6**) reflects the stability of the incipient benzyl radical, and is consistent with a one-step mechanism [Equation (1)] in which the decarboxylation is concerted with fission of the peroxide bond. The parent compound, on the other hand, is believed to give methyl radicals in a stepwise process [Equation (2)].[1] Support for this comes from a more detailed kinetic study which shows that the two reactions have quite different Arrhenius

$$PhCH_2C(=O)O{-}OBu^t \ (\mathbf{5}) \longrightarrow PhCH_2\cdot + CO_2 + \cdot OBu^t \qquad (1)$$

[1]Remember (Chapter 2) that methyl radicals can also arise by fragmentation of the butoxyl radical.

parameters; the activation entropies are consistent with a more highly ordered transition state for the phenylperoxyacetate ($\Delta S^{\ddagger}$ = +2 cal deg^{-1} mol^{-1} vs. +20 cal deg^{-1} mol^{-1} for the parent peroxyacetate).

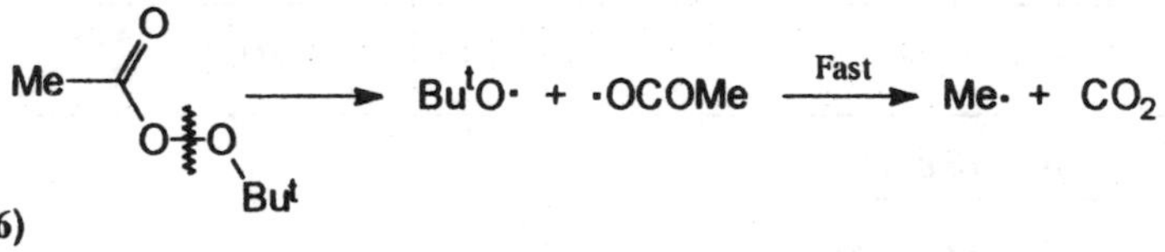

$$\text{(6)} \longrightarrow Bu^tO\cdot + \cdot OCOMe \xrightarrow{\text{Fast}} Me\cdot + CO_2 \quad (2)$$

The peroxyoxalate (7), a useful (but notoriously hazardous) source of butoxyl radicals having a half-life of only a few minutes at 40°C, is believed to give t-butoxyl radicals by a process in which three bonds rupture simultaneously.

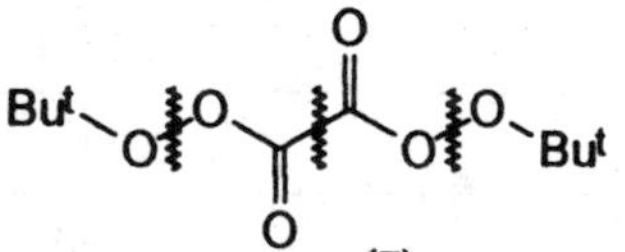

(7) 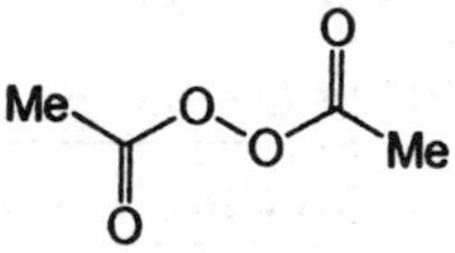 (8)

Even with the simple peroxyacetate, or with diacetyl peroxide (**8**), the acetoxyl radicals are so unstable towards decarboxylation that it is generally considered that any reaction products containing the acetoxyl function are likely to have arisen by some mechanism which by-passes free acetoxyl radicals. However, σ-bonds to sp^2-hybridised carbon are generally stronger than those to tetrahedral carbon, so that benzoyloxyl radicals from dibenzoyl peroxide (**9**) do sometimes give products which incorporate benzoate groups (e.g. Scheme 3.1). This has also been observed for a few rather exceptional alkanoyloxyl radicals, such as cyclopropanecarboxyl (**10**).

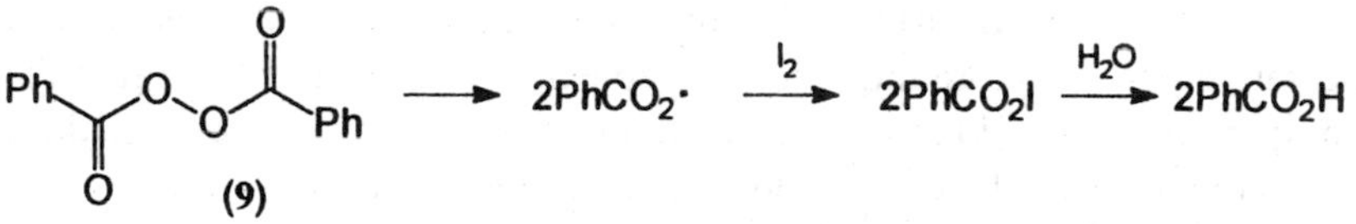

$$\text{(9)} \longrightarrow 2PhCO_2\cdot \xrightarrow{I_2} 2PhCO_2I \xrightarrow{H_2O} 2PhCO_2H$$

Scheme 3.1: Thermolysis of benzoyl peroxide (dibenzoyl peroxide) in a wet (H_2O) solution of iodine in CCl_4 gives an almost quantitative yield (2 mol/mol) of benzoic acid. The rate of reaction is that expected of the unimolecular homolysis of the peroxide.

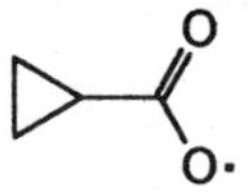

(10)

Whilst many carboxyl radicals decarboxylate extremely readily, the much slower fragmentation of the t-butoxyl radical has frequently been used to compare rates of

1 cal = 4.184 J

hydrogen abstraction from various substrates by this radical. Making the assumption that the fragmentation of butoxyl radicals is independent of solvent (which seems to be reasonably justified at least for non-polar solvents), the ratio of t-butanol to acetone in the products is indicative of the relative rates of hydrogen abstraction by, and fragmentation of, the t-butoxyl radical under any particular set of experimental conditions (Scheme 3.2).

$$Me_3C\text{–}O\cdot \xrightarrow[k_{abstr}]{R\text{-}H} Me_3C\text{–}OH + R\cdot$$

$$Me_3C\text{–}O\cdot \xrightarrow{k_{frag}} Me_2CO + Me\cdot$$

Scheme 3.2

In the laboratory, peroxides must always be handled with respect. As a rule of thumb, the greater the proportion of "non-peroxide mass" to "peroxide mass" the less hazardous is the compound. Associated with molecules like acetyl peroxide and di-t-butyl peroxyoxalate (as well as some of the peroxides derived from autoxidation of ethers), where the peroxide oxygens comprise a significant fraction of the molecule, there is a very real explosion risk. Such compounds must be treated with extreme caution.

Even in solution, reactions of organic peroxides may be violent, because some of these compounds, particularly the diacyl peroxides, are especially prone to induced decomposition; i.e. as well as unimolecular dissociation, they may react in bimolecular homolytic processes in which the second reactant may be anything from an apparently innocuous molecule of solvent to a reactive radical generated as a consequence of the unimolecular decay. Examples are discussed later in this chapter, as well as in Chapter 9.

"Caged" versus "free" radicals: When a solution of AIBN **(3)** in oxygen-free benzene is heated, the only products are dimers of the cyanopropyl radical **(4)**. These are principally **(11)**, but also some **(12)**, indicative of reaction in the resonance form **(4')**, although **(12)** is also unstable and itself dissociates into cyanopropyl radicals not much more slowly than does AIBN. If the experiment is

$$\textbf{(3)} \longrightarrow 2\ NC\text{–}\dot{C}Me_2\ \textbf{(4)} \longleftrightarrow \cdot N{=}C{=}CMe_2\ \textbf{(4')}$$

$$\textbf{(4)} \rightarrow NC\text{–}CMe_2\text{–}CMe_2\text{–}CN\ \textbf{(11)}$$

$$\textbf{(4)} \rightleftharpoons NC\text{–}CMe_2\text{–}N{=}C{=}CMe_2\ \textbf{(12)}$$

repeated, but present in the system there is a quantity of a stable radical, e.g. the purple diphenyl picryl hydrazyl [DPPH; (**13**)], or a reactive alkene, such as styrene, the yield of cyanopropyl radical dimers is reduced, since some will react with the additive before they can dimerise. A plot of the yield of dimer against the quantity of additive assumes the general form of Fig. **3.1**. From the graph, it is evident that not all of the cyanopropyl radicals can be successfully intercepted, however much additive is present. This intriguing result is a consequence of the "cage effect".

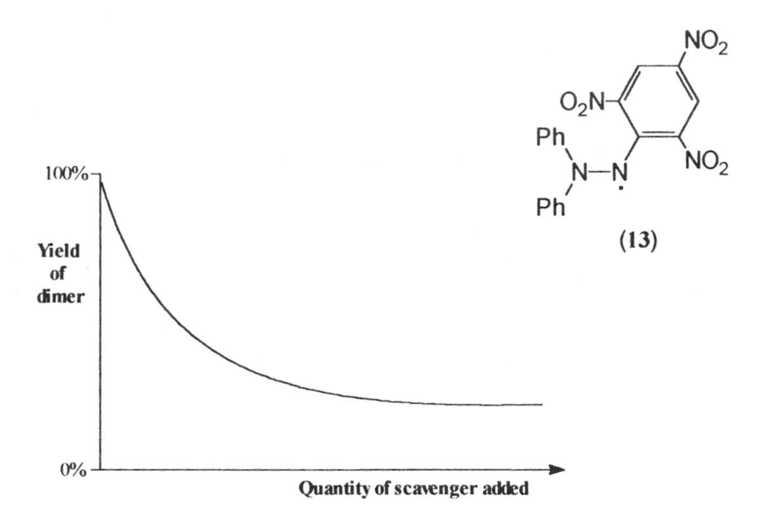

Fig. 3.1: The decomposition of AIBN (**3**) in benzene. Schematic representation of the variation in yield of 2,3-dicyano-2,3-dimethylbutane as a function of the concentration of added radical scavenger.

When a molecule of AIBN decomposes then two cyanopropyl radicals are generated in close proximity so that there is a significant probability that the two will collide before they have a chance to diffuse away from one another. Since dimerisation is an encounter-controlled, activationless process, those radicals which do collide will very probably interact. It is said that, by virtue of not diffusing apart, they have not "escaped from the solvent cage". Dimerisation under these circumstances is referred to as "geminate" recombination: prior to combination or to diffusive separation, the two radicals constitute a "geminate pair". Radicals which do not escape the solvent cage ("caged" radicals) are so short-lived that they cannot participate in radical-molecule reactions which are orders of magnitude too slow. Only those radicals which do diffuse apart – to become "free" radicals – are sufficiently long-lived to engage in radical-molecule reactions such as addition to styrene (or even trapping

1 cal = 4.184 J

by a stable radical like DPPH[1]). However, in the absence of styrene, or of some other additive, the free cyanopropyl radicals will eventually re-encounter and dimerise, giving product indistinguishable from the cage product. This behaviour is elaborated in Scheme 3.3. The limiting yield of dimer in Fig. **3.1** gives an indication of the magnitude of the cage effect in AIBN decomposition. When azo-compounds or peroxides are being used as initiators for chain reactions, it will be clear that only the *free* radicals from the decomposition can be effective; "initiator efficiency" is therefore often less than unity.

$$R{-}N{=}N{-}R \longrightarrow [R\cdot + N_2 + R\cdot]_{\text{Solvent cage}}$$

Diffusive separation → $N_2 + 2R\cdot$ ("Free" radicals) $\longrightarrow$ R—R

Cage recombination → R—R

Scheme 3.3: A simlified schematic representation of the cage effect. The "solvent cage" surrounding the pair of radicals immediately following azo-compound homolysis is represented by square brackets.

Not surprisingly, the proportion of cage recombination may be increased by using viscous solvents. There is also evidence that the extent of cage recombination is limited by the "spacer" molecules such as nitrogen or carbon dioxide that separate the radical pair immediately the initiator molecule dissociates.

For our purposes, this represents an adequate introduction to the cage effect. More subtle features, such as a distinction between "primary" and "secondary" radical pairs, are beyond the scope of the present discussion. Nevertheless, there is one other important aspect of radical pair behaviour to which we must return later (Chapter 5).

Polar effects in initiator thermolysis: Although most of the radicals which we have encountered so far are electrically neutral, there are many circumstances in which the rates of their reactions, or of reactions in which they are formed, are clearly influenced by polar effects. A rationalisation of the rate differences exhibited between the (concerted two-bond) decompositions of (**5**) and its *p*-methoxy- and *p*-nitro-derivatives (Table 3.2) provides us with a first insight into these effects in radical chemistry. In the incipient butoxyl radical from each decomposition, the unpaired electron becomes localised on a highly electronegative oxygen atom –

[1]Unless the stable radical is present at very high concentration, under which circumstances it may effectively be present in the "walls of the cage".

capable of stabilising negative charge. It is nowadays generally accepted that the transition state for decomposition of **(5)**, which can be represented by **(14)**, incorporates a modest contribution from the resonance structure **(14')**.[1] In this structure, the benzyl cation would be stabilised by the *para*-methoxy-substituent but destabilised by the *para*-nitro-substituent. The additional stabilisation to the transition state which is imparted by the methoxy-group results in the greater rate of decomposition of the corresponding peroxide. The reverse is true for the nitro-derivative.[2]

$PhCH_2$----C(=O)---O----O-Bu^t ⟷ $PhCH_2^+$ CO_2 ^-O-Bu^t

(14) **(14')**

Table 3.1 showed some rate data for azo-compound decomposition. Secondary kinetic isotope effects on the rates of mono- and di-α-deuterated derivatives of the symmetrical α-phenylethylazo-compound **(15)** indicate equivalent rate effects at both ends of the molecule, consistent with the symmetrical nature of the decomposition transition state. Likewise, substituent effects on the rate are attributable to radical-stabilising factors rather than to polar ones which might be expected if the transition states were unsymmetrical, as necessitated by single-bond fission. Single-bond fission may, however, be the rate-limiting step for the decomposition of highly unsymmetrical azo-compounds such as phenylazotriphenylmethane **(16)**.

PhC(Me)(H(D))–N=N–C(Me)(H(D))Ph Ph_3C–N=N–Ph

(15) **(16)**

[1]The transition state will have more diffuse bonding, and be more polarisable, than the initial peroxide. Consequently, the unpairing of electrons and the shift of electronic charge as individual bonds break are not perfectly correlated with the stretching of those bonds.

[2]Whilst the prime objective of this section is to introduce the importance of polar effects in *homolytic* reactions, it has neglected another important aspect of peroxide decomposition, namely that many peroxides may also decompose by ionic pathways. These are particularly evident in polar solvents, and with peroxides which are markedly unsymmetrical (e.g. p-$MeOC_6H_4CO.OO.COC_6H_4NO_2$-$p$).

1 cal = 4.184 J

3.2 PHOTOLYSIS

A quantum of ultraviolet radiation, or even of visible light, carries sufficient energy to break many different bond types in organic compounds. For example a quantum of visible light[1] of wavelength λ = 500 nm is equivalent to approximately 57 kcal mol^{-1}, and this is more than sufficient to split di-t-butyl peroxide, for which we have seen that dissociation of the peroxide bond requires only about 38 kcal mol^{-1}. Nevertheless, di-t-butyl peroxide is completely stable towards irradiation at this wavelength – for the simple reason that the peroxide is wholly transparent at 500 nm. The so-called "First Law of Photochemistry" reminds us that in order for light to bring about photochemical change, it *must* first be absorbed! Di-t-butyl peroxide is almost colourless, but absorbs light in the near ultraviolet and this does promote efficient photodissociation into butoxy radicals. All peroxides, as well as azo-compounds, are easily photolysed by ultraviolet light. This includes alkyl hydroperoxides and hydrogen peroxide, which are a little more stable than other organic peroxides towards simple thermolysis. Specifically, in the case of hydrogen peroxide, photolysis affords a means of generating hydroxyl radicals.

Amongst other examples of photochemical radical production is the photolysis of aryl iodides. This affords a particularly convenient method of generating phenyl and other aryl radicals. The production of 3-hydroxybiphenyl in good yield by ultraviolet irradiation of a benzene solution of *m*-iodophenol (Scheme 3.4) not only illustrates the utility of such photolyses, but also exemplifies the fact that radical reactions are frequently unaffected by the presence of functionality which might have to be protected before carrying out reactions involving ionic intermediates.

HO–C₆H₄–I $\xrightarrow{h\nu}$ HO–C₆H₄· + I· (⟶ I_2)

PhH

HO–C₆H₄–C₆H₆· (H) $\xrightarrow{\text{I· or } I_2}$ HO–C₆H₄–C₆H₅ + HI

Scheme 3.4

3.3 RADIOLYSIS

Interaction of matter with "ionising radiation", i.e. high-energy electromagnetic radiation (X- or γ-rays) or with α- or β-particles, can promote chemical change which commonly involves free radicals. With quanta having X- or γ-ray wavelengths, initial interaction with molecules releases fast electrons which

[1]Strictly, a mole of quanta!

themselves each carry sufficient energy to further disrupt hundreds or even thousands of additional molecules. Fast electrons, in the form of β-particles can cause similar molecular damage. Principally, this involves ejection of an outer shell electron, with some of the resulting "secondary electrons" being sufficiently energetic to ionise yet more molecules.

These brief introductory remarks apply to radiolysis of all simple molecular species. However, one of the major areas of concern has been radiation damage to the largely aqueous medium of living tissues, where water molecules are ionised to give H_2O^+, and subsequent reaction of this with H_2O forms H_3O^+ and the very reactive HO· radical. In addition, combination of H_2O^+ with an electron gives an excited water molecule which is sufficiently energetic to dissociate into HO· and H·. For more detailed discussion the interested reader is referred to the bibliography, but it should already be clear that ionising radiation can produce a cascade of reactive radicals.

3.4 ELECTRON TRANSFER AND MOLECULE-INDUCED HOMOLYSIS

A classic transformation in radical chemistry is the Fenton reaction between iron(II) salts and hydrogen peroxide. In an extremely rapid process, iron(II) is oxidised to iron(III), whilst the peroxide is reduced, dissociating into hydroxide and a hydroxyl radical [Equation (3)].

$$Fe^{++} + H_2O_2 \longrightarrow Fe^{+++} + HO^- + HO\cdot \qquad (3)$$

The majority of induced peroxide decompositions fit a pattern of electron transfer, with rates which depend on the redox potential of the species which induces the decomposition. Transition-metal ions are obvious candidates, but the pattern is also evident when the peroxide decomposition is induced by a reactive radical or by a fully bonded organic. For example when dibenzoyl peroxide decomposes in a dialkyl ether, the reaction is appreciably faster than in an alkane at the same temperature. This is because in the ether, alkoxyalkyl radicals are formed and these are sufficiently easily oxidised that some of them induce the decomposition of the peroxide (Scheme 3.5). The same peroxide should not be added to *N,N*-dimethylaniline, since the latter is so easily oxidised that a vigorous peroxide decomposition is promoted, apparently by the amine itself rather than by derived radicals. This last example constitutes an instance of "molecule-induced homolysis".

Many reactions of diazonium salts are radical in nature. One instance of this was revealed with the discovery that attempted synthesis of the aryl iodide (**19**), using the conventional reaction of the diazonium salt (**17**) with iodide, gave instead the cyclised product (**20**) (Scheme 3.6). The intention had actually been to synthesise this compound, but by photolysis of the expected iodide. Further

1 cal = 4.184 J

investigations of this type of reaction have established quite unequivocally that the iodide-induced cyclisation does involve an intermediate aryl radical **(18)**.

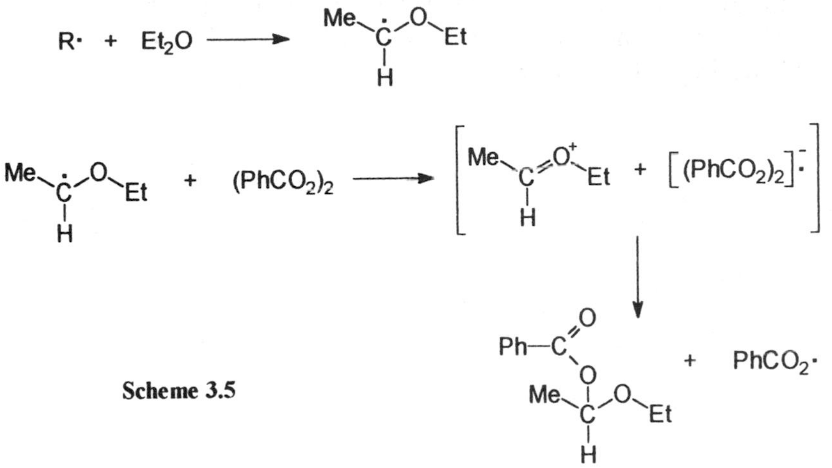

Scheme 3.5

Such a result raises the interesting question of the mechanism of diazonium displacement by iodide when there is no neighbouring aryl group available for intramolecular attack. It seems likely that an intermediate aryl radical is intercepted by the iodide ion to give initially the radical anion, ArI$^{\cdot-}$. This type of behaviour will be discussed further in Chapter 8. Whether the corresponding copper-promoted

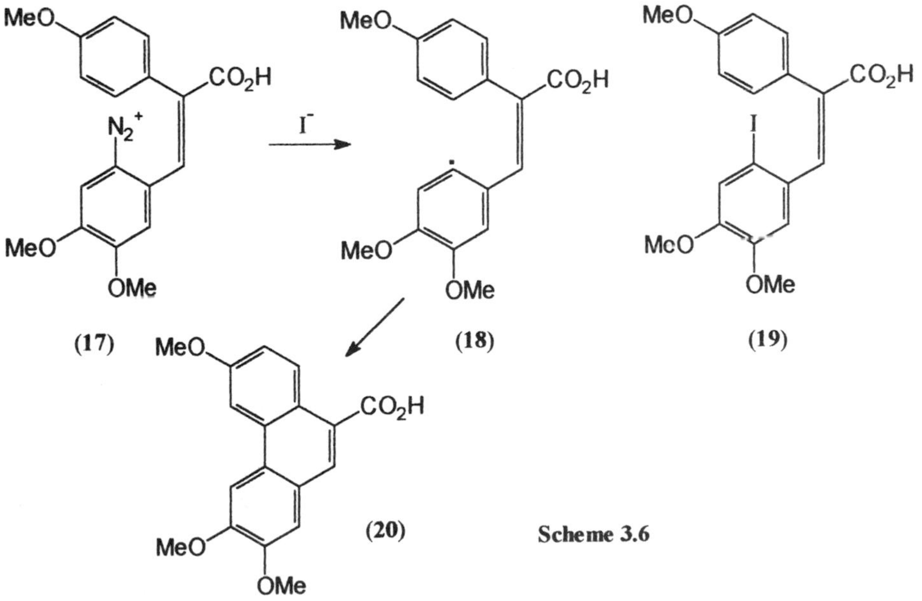

(17) (18) (19) (20)

Scheme 3.6

preparations of aryl bromides and chlorides are similar, or involve arylcopper species (instead or as well), remains unclear.

By way of conclusion, it is worth remembering that valuable probes for some of these redox processes are available in the armoury of the electrochemist; and, as we have seen in recalling the Kolbe reaction (Chapter 2), radicals are familiar intermediates in electrosynthesis.

1 cal = 4.184 J

4

An introduction to reactivity and selectivity in radical reactions

4.1 INTRODUCTION

The discussion in the previous three chapters has given some insight into the *nature* of radical reactions. It will also have become clear that not all radicals are likely to participate in all of these. Thus we have encountered species ranging from triphenylmethyl which, in suitable (oxygen-free) solvents such as benzene, is in stable equilibrium with dimer, to $CH_3\cdot$, which is so reactive that it will rapidly attack even this relatively resistant substrate. The benzyl radical, $PhCH_2\cdot$ (Fig. **4.1**), enjoys less steric congestion round the principal radical centre and somewhat less resonance stabilisation than $Ph_3C\cdot$ (see Chapter 1). When the benzyl radical is formed in benzene it is not sufficiently reactive to form product by attacking the solvent, but it is rapidly destroyed by *irreversible* dimerisation. Evidently we are dealing with three related radicals which exhibit pronounced differences in reactivity under similar experimental circumstances.

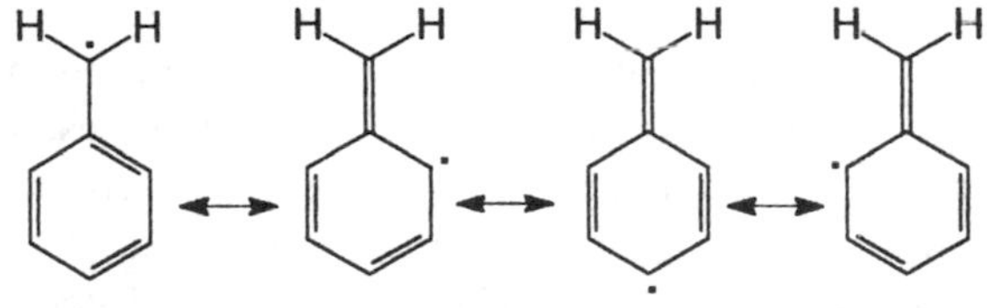

Fig. 4.1: Resonance structures for the benzyl radical; the principal contributor is on the left.

In a comparable fashion, it is easy to find examples in which the reactivities of different solvents vary markedly towards a given radical. For example, if trichloromethyl radicals are generated in benzene they behave like benzyl and dimerise (to form hexachloroethane). On the other hand, in toluene, a major fate is reaction with solvent by hydrogen abstraction from the methyl group.

The object of this chapter is to attempt to explain the various factors which influence these patterns of reactivity, and to develop a semi-quantitative understanding of them. In many respects it constitutes the most important section of this book.

4.2 HALOGENATION OF ALKANES

We shall begin our analysis with the halogenation of alkanes. The propagation steps for reaction with 2-methylpropane (isobutane) are given in Scheme 4.1 (where X represents a chlorine or a bromine atom). It should be evident that the ratio of tertiary halide to primary halide is determined by the competition between steps **ia** and **ib**, i.e. between abstraction of tertiary and primary hydrogen.[1]

$$(CH_3)_3C{-}H + X\cdot \xrightarrow{\textbf{ia}} (CH_3)_3C\cdot + HX \xrightarrow{X_2} (CH_3)_3C{-}X + X\cdot$$

$$(CH_3)_3C{-}H + X\cdot \xrightarrow{\textbf{ib}} HX + (CH_3)_2CH{-}CH_2\cdot \xrightarrow{X_2} (CH_3)_2CH{-}CH_2X + X\cdot$$

Scheme 4.1

The observed results for the two halogens are very different, as indicated in Table 4.1. It is possible to rationalise this difference in behaviour simply by saying that the chlorine atom is very much more reactive than the bromine atom and is therefore less selective in its reactions. However, a more useful description is obtained by considering reaction profiles for the hydrogen transfer processes. Before proceeding with that, it is very important to recognise that the near equality of yields of primary and tertiary *chlorides* shown in Table 4.1 does *not* indicate that chlorine is almost completely unselective. The yields of primary and tertiary chlorides are determined by the relative reactivities of the two sites *coupled with a statistical factor* – there are nine equivalent primary hydrogens in each molecule of 2-methylpropane but there is only one tertiary hydrogen. This seemingly obvious consideration is all too easily overlooked. Comparisons of reactivity are more

[1]Strictly, this requires that the steps are irreversible, and also that the two alkyl radicals cannot directly interconvert.

1 cal = 4.184 J

usefully made on a "per hydrogen atom" basis. Recognising this, we can present the data of Table 4.1 as shown in Rows 2 and 4 of Table 4.2. The *relative* reactivity of primary hydrogen is taken as unity for both halogens. The data in Table 4.2 represent averages of results obtained using several different aliphatic hydrocarbons. From these numbers one can predict that, from 2-methylpropane, primary and tertiary chlorides should be obtained in the proportions 9 : 5.5.

Table 4.1: Products from halogenation of 2-methylpropane.

	$(CH_3)_3CX$	$(CH_3)_2CHCH_2X$
Chlorination (X = Cl)	ca. 40%	ca. 60%
Bromination (X = Br)	> 99%	< 1%

The kinetics and energetics of abstraction of hydrogen atoms from saturated hydrocarbons by halogen atoms have been well studied, both in the gas phase and in solution. The reaction of a chlorine atom with methane is illustrative. Dissociation of a C–H bond in methane requires some 105 kcal mol^{-1}, whilst the dissociation energy of HCl is 103 kcal mol^{-1} (see Table 4.3). In other words, the reaction: $Cl\cdot + CH_4 \rightarrow HCl + CH_3\cdot$ is endothermic, but by only 2 kcal mol^{-1}. A remarkable feature of this process is the extraordinary compensation which occurs between the energy of the bond being broken and that of the bond being formed, so that the transition state, (**1**), which has a three-centre bond embracing chlorine, hydrogen and carbon, is no more than 4 kcal mol^{-1} above the energy of the reactants (Fig. **4.2** overleaf). (It is pertinent to note here that molecular orbital calculations confirm that in the transition state the three key atoms are co-linear.)

Table 4.2: Relative reactivities "per hydrogen" of saturated hydrocarbons towards Cl· and Br·.

	Cl·	Br·
1: CH_3–H	0.003	–
2: RCH_2–H	1.0	1.0
3: R_2CH–H	4	82
4: R_3C–H	5.5	ca. 1600

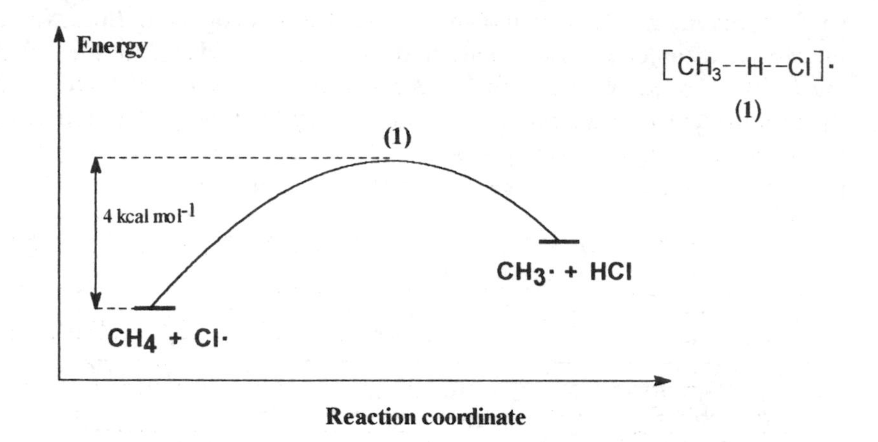

Fig. 4.2: Reaction profile for hydrogen abstraction from methane by a chlorine atom.

Other hydrogen transfers to halogen can be related to Fig. **4.2** simply by "skewing" the reaction profile according to the thermodynamics of the reaction. For example, in an exothermic process the right-hand side of Fig. **4.2** will be pulled downwards relative to the left-hand side; the transition state will be pulled down to a lesser extent and will drift to the left (Fig. **4.3**). This is conveniently illustrated by considering in more detail the chlorination and bromination of 2-methylpropane.

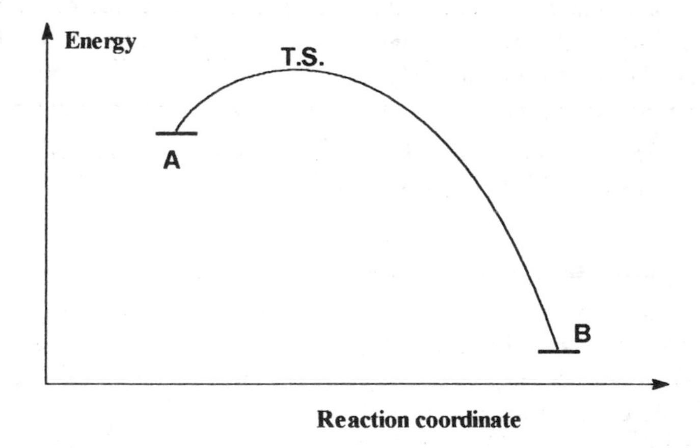

Fig. 4.3: Generalised reaction profile for an exothermic reaction step. Note that the transition state (**T.S.**) is closer to the reactants (**A**) than to the products (**B**) along *both* coordinate axes.

The gas-phase bond-dissociation energies (b.d.e.s) of H–Br (87.5 kcal mol^{-1}) and H–Cl (see above) are known with some precision from spectroscopic studies.

1 cal = 4.184 J

Similar accuracy is not available for the dissociation of C–H bonds in alkanes (except methane), and values have been revised several times in the last decade. A selection of b.d.e.s is collected in Table 4.3. Possible errors of 1 kcal mol^{-1} or more may be present in many of the figures – though it is improbable that the relative order of b.d.e. values, CH_3–H > RCH_2–H > R_2CH–H > R_3C–H, will need revising.

Table 4.3: Dissociation energies of selected bonds (kcal mol^{-1} at 300 K).*

Bond		Bond		Bond	
CH_3–H	105	Cl_3C–H	96	$((CH_3)_3Si)_3$Si–H	79
CH_3CH_2–H	101	F–H	136	$(CH_3)_3$Sn–H	74
$(CH_3)_2$CH–H	98	Cl–H	103	C_2H_5–Cl	81
$(CH_3)_3$C–H	95	Br–H	87.5	C_2H_5–Br	69
CH_2=CH–H	103	I–H	71	C_2H_5–I	53
HC≡C–H	130	HO–H	119	RO–OR	ca.37
C_6H_5–H	111	HOO–H	88	CH_3–CH_3	89
CH_2=$CHCH_2$–H	87	CH_3O–H	105	CH_3CH_2–CH_3	87
$C_6H_5CH_2$–H	89	C_6H_5O–H	86	$(CH_3)_2$CH–CH_3	86
RC(=O)–H	87	R_2NO–H	ca.74	$(CH_3)_3$C–CH_3	84
$C_2H_5OCH(CH_3)$–H	92	CH_3S–H	92	Cl–Cl	58
N≡CCH_2–H	86	C_6H_5S–H	82	Br–Br	46
CH_3COCH_2–H	92	$(CH_3)_3$Si–H	90	I–I	36

***The figures represent gas-phase enthalpies of bond homolysis.**

This is consistent with a radical stability order tertiary > secondary > primary > methyl which may be associated with a stabilising effect of alkyl substitution, discussed in Chapter 5. Using the figures of Table 4.3, and taking first the *chlorination* of 2-methylpropane, it is apparent that abstraction of hydrogen from either site will be exothermic, and that the reaction coordinate diagrams will both take the form of Fig. **4.3**. This is illustrated in Fig. **4.4**. From the foregoing discussion, the activation energies for both processes are predicted to be less than that for reaction with methane, and the difference between them will be very small. By attempting to draw similar curves, one might estimate that difference as very approximately 1 kcal mol^{-1}. Now, if we make the further assumption that the pre-exponential factors in the Arrhenius equations that govern the two reactions are the same, then at ca. 300 K this corresponds to a rate difference (per hydrogen) of a factor of ca. 5 (the reader is invited to check this). The experimental value of 5.5 (Table 4.2) is reassuringly close.

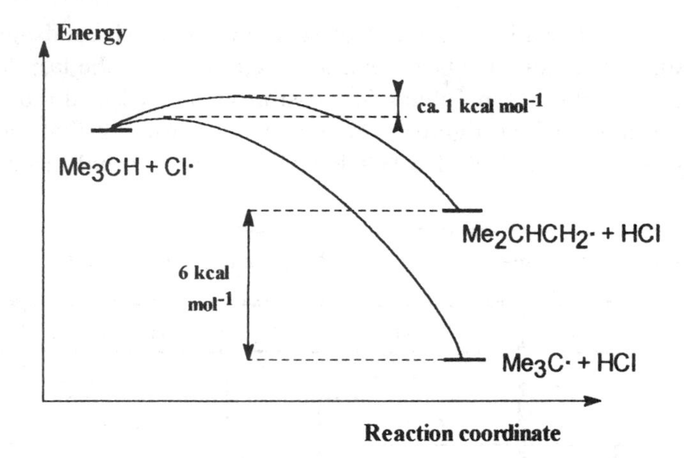

Fig. 4.4: Reaction profiles for hydrogen abstraction from 2-methylpropane by chlorine atoms.

Bromination can be treated similarly, except that both hydrogen transfers are now endothermic and we arrive at reaction profiles such as those illustrated in Fig. **4.5**.

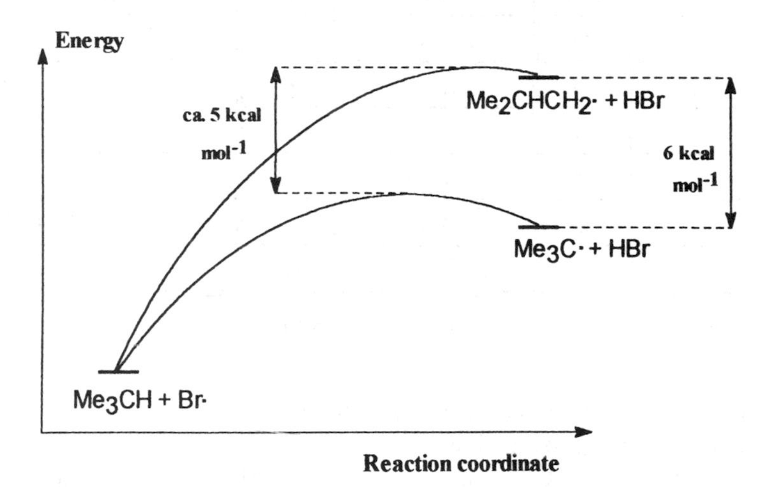

Fig. 4.5: Reaction profiles for hydrogen abstraction from 2-methylpropane by bromine atoms.

Examination of Fig. **4.5** shows quite clearly that the reverse reactions must have very much higher rate constants than the forward ones. Indeed, the activation barriers for the reverse reactions will be very small indeed. However, under normal reaction conditions, the interaction of an alkyl radical with Br_2 is strongly exothermic and essentially free from an activation barrier. At the commencement of reaction, the concentration of bromine will far exceed that of HBr, so that the

1 cal = 4.184 J

only important reactions of the alkyl radicals will be with Br_2. It is also clear from the figure that the difference in activation energies for the endothermic hydrogen-transfers is only marginally less than the difference between the endothermicities of the two reactions: say 5 kcal mol^{-1} compared with 6 kcal mol^{-1}. Using an analysis similar to that for chlorination, this translates into a per-hydrogen reactivity ratio at 300 K of approximately 5000, only a little larger than the experimental result given in Table 4.2.

These analyses embrace a principle in chemistry usually referred to as the Hammond Postulate. This can be presented in various ways, but is simply stated as follows: "The transition states of exothermic reaction steps are generally reactant-like, whilst those of endothermic reaction steps are generally product-like". Thus for chlorination of 2-methylpropane, the transition states for the hydrogen transfer steps are found "early" on the reaction coordinate, and are close to the reactants both in energy (Fig. **4.4**), and in geometry [structure **(2)**]. In contrast, for the bromination reaction the transition states are found "late" on the reaction coordinate, close to the products [Fig. **4.5** and structure **(3)**].

[—C- -H-------Cl]·   [—C-------H- -Br]·

(2)   **(3)**

4.3 OTHER REACTIONS

When comparing any pairs or sets of similar reactions, the type of analysis presented above often provides a reasonable guide to relative reactivities. For example, any given radical, X·, might be expected to add less readily to ethene than to styrene or to butadiene, both of which form resonance-stabilised adduct radicals.

Included amongst very extensive studies of radical additions have been detailed calculations of reaction trajectories – using molecular orbital (MO) theory at various levels of sophistication. For example the approach of a methyl radical to ethene has been found to pass through a transition state with the geometry shown in Fig. **4.6.**

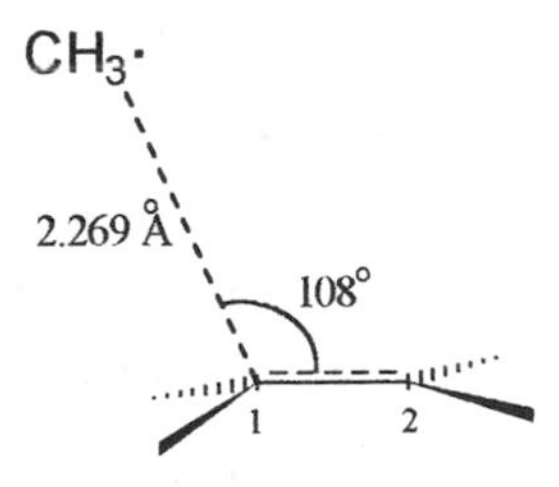

Fig. 4.6: Calculated transition state for the addition of a methyl radical to ethene. Note the concurrent displacement of substituents on the ethene as the hybridisation at C-1 changes from sp^2 to sp^3. Interestingly, there is also a slight upwards displacement of substituents at C-2.

At the highest *ab initio* level investigated, the calculations indicate an appreciable activation barrier of ca. 8.4 kcal mol^{-1}. This is very close to the experimental value, although in the light of the earlier analysis the barrier may seem somewhat higher than expected since the reaction is quite exothermic (ca. 23 kcal mol^{-1}). Important consequences of this barrier will be elaborated later.

A simple frontier orbital picture of the approach geometry (Chapter 9) for methyl addition to ethene views the *p*-electron of the (planar[1]) methyl radical feeding into the vacant π*-orbital of the hydrocarbon (Fig. **4.7**), although a full description must take into account the relocation of the bonding electrons.

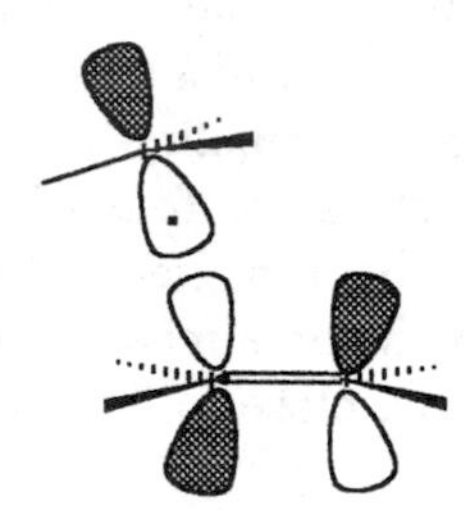

Fig. 4.7: An orbital picture of the approach of the methyl radical *p*-orbital to the π*-orbital of ethene.

The results of calculations, as well as much of the experimental data, refer to the gas phase. But as already pointed out, except where electron transfer is involved, radical chemistry is simplified by the usually small magnitude of any solvent effect. There are, however, other important effects, and these may operate in such a way as to over-ride the generalisations given above. The results of these can be expressed in reaction profile diagrams such as Fig. **4.8**, where the graphs for two superficially similar processes are depicted as "crossing". In this illustration the

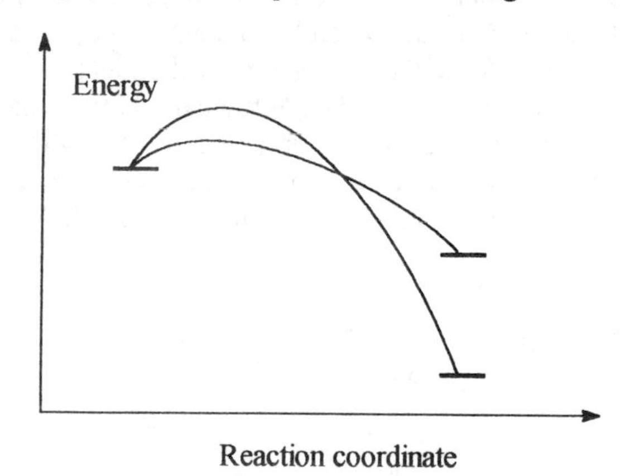

Fig. 4.8: "Crossing" reaction profiles.

[1]Evidence for the planarity of the radical centre in simple alkyl radicals will be discussed in the next chapter.

1 cal = 4.184 J

more exothermic process has the higher activation barrier. From the examples discussed below and in subsequent chapters it will be seen that these other influences on reaction rate may be very important indeed.

4.4 STEREOELECTRONIC EFFECTS

We noted in Chapter 2 that the 5-hexenyl radical cyclises predominantly to the cyclopentylmethyl radical rather than to the isomeric cyclohexyl. A moment's reflection should suggest that this must be an example of "crossing" reaction profiles, since (a) cyclopentylmethyl is primary whilst cyclohexyl is secondary, and (b) the 5-membered cycloalkane ring is generally more strained than is the 6-

ca. 3% ← $CH_2\cdot$ → ca. 97% $CH_2\cdot$

membered. There is, however, a special factor operating in the case of cyclohexyl formation: at the transition state there is appreciable ring-strain imposed by the geometric requirements for orbital overlap of the kind illustrated in Fig. **4.7**. Inspection of a suitable model shows this very clearly. It is much easier to position the planar radical centre of the hexenyl species so that it can interact with the π^*-orbital at C-5 than it is for the interaction to be at C-6; for the latter case some strain is quite evident in the model.

Usually, 5-membered-ring formation involves a chair-like transition state (Fig. **4.9a**), but when this is raised in energy by unfavourable interactions between substituents, a boat-like alternative is also accessible (Fig. **4.9b**). This cyclisation has assumed considerable synthetic importance (Chapter 6), and has received correspondingly detailed theoretical scrutiny. The arguments based on simple

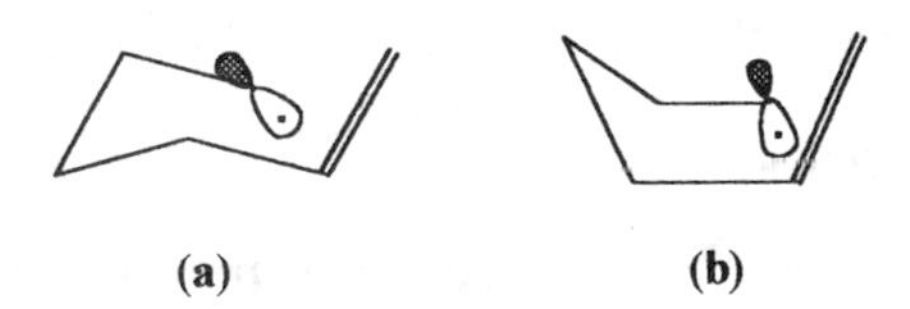

Fig. 4.9: Representations of chair-like and boat-like transition states for hexenyl radical cyclisation.

overlap considerations are fully borne out both by semi-empirical molecular orbital calculations, and by molecular mechanics computations on the possible transition states. The latter calculations depend on force-field parameters for the transition state developed from the *ab initio* MO analysis of methyl radical addition to ethene mentioned above. The importance of favourable orbital overlap at the transition state, which in this example steers the reaction towards the thermodynamically less stable cyclopentylmethyl radical, is commonly referred to as a "stereoelectronic effect" (although some chemists prefer to reserve this term for special lone-pair effects in reactions of carbohydrates and related molecules).

A second example in which stereoelectronic factors control a free-radical reaction is the rearrangement of a cyclopropyl radical into an allyl radical. Such a process would be highly exothermic (ca. 30 kcal mol^{-1}), yet the activation barrier (ca. 19 kcal mol^{-1} measured in gas-phase studies carried out at elevated temperature) is too high for it to occur in solution in the life-time of a cyclopropyl radical. Only when additional features are present which impart substantial extra stabilisation to the developing allyl radical, such as two phenyl groups, is the rearrangement observed in solution. The problem here is that the rearrangement has to be well developed before any allylic stabilisation is experienced, resulting in the observed high activation barrier.

CH·

Ph CH· Ph Ph Ph

Ph Ph CH· Ph Ph

Ring-opening of a cyclopropyl radical may be regarded as an example of a pericyclic radical rearrangement. Calculations have indicated that there should be no dramatic preference for conrotatory or disrotatory opening. There is as yet no unambiguous experimental evidence on this point. Examples of the more extensive studies carried out on pericyclic reactions of radical ions are given in Chapter 8.

4.5 POLAR INFLUENCES

A completely different, and very important transition-state effect is illustrated by consideration of hydrogen abstraction from methyl acetate by t-butoxyl radicals. In methyl acetate the weaker C–H bonds are those in the acetyl group (although the figures are a little uncertain, the difference is probably at least 5 kcal mol^{-1}), yet it is the other methyl group which is preferentially attacked. The pictorial rationalisation of this is in terms of a polar effect in which the transition state may be stabilised by partial charge separation. The electronegative oxygen of the butoxyl radical polarises the transition state for hydrogen transfer as indicated in

1 cal = 4.184 J

(**4**).[1] This will be assisted if there are any substituents present which will stabilise the partial positive charge on carbon. Conversely, any electron-withdrawing substituents on carbon will make the reaction more difficult. In the case of methyl acetate, it should be evident from structures (**5**) and (**6**) that the oxygen lone-pair will facilitate abstraction of the methoxyl hydrogen, but that the electron-withdrawing carbonyl will inhibit reaction at acetyl hydrogen.

$$[RCH_2\text{---}H\text{---}OBu^t]^{\cdot} \longleftrightarrow [RCH_2^{+}\;\;\dot{H}\;\;{}^{-}OBu^t]$$

(**4**)

$$\left[\begin{array}{l}MeCO\\ \quad\;\; O^{+}{=}CH_2 \;\;\; \dot{H} \;\;\; {}^{-}OBu^t\end{array}\right] \quad \left[\begin{array}{c}O^{\delta-}\\ \| \\ C^{\delta+}\\ MeO \diagup \quad \diagdown CH_2^{+} \;\;\; \dot{H} \;\;\; {}^{-}OBu^t\end{array}\right]$$

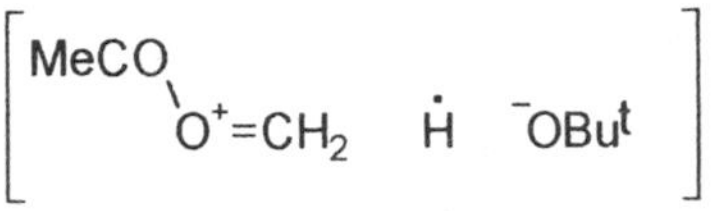

(**5**) 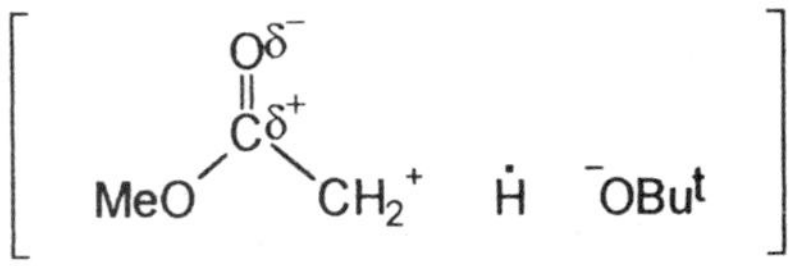 (**6**)

As we have already noted for peroxyester decompositions (Chapter 3), in those processes in which polar factors affect the rates of radical reactions, the movement of electronic charge is not perfectly synchronised with bond-breaking and bond-making.

Some well-known examples of polar influences in hydrogen abstraction are found in ether peroxidations. These reactions are often complex, but are dependent on autoxidation (see Chapter 2) adjacent to ether oxygen. The ease of hydrogen abstraction at this site by alkylperoxyl radicals seems to be significantly greater than might be expected solely on the basis of any bond-weakening due to the ether oxygen. The dipolar transition state is here depicted in (**7**). The derived hydroperoxides are usually involved in further reactions. For instance, the dangerously explosive peroxide which crystallises from diisopropyl ether on prolonged exposure to air and light has the structure (**8**).

$$\left[RO{-}\overset{|}{\underset{|}{C}}\text{-----}H\text{-----}OOR'\right]^{\cdot} \longleftrightarrow \left[RO{-}\overset{|}{\underset{|}{C}}{}^{+} \;\;\; \dot{H} \;\;\; {}^{-}OOR'\right]$$

(7)

$$(CH_3)_2CH{-}O{-}CH(CH_3)_2 \longrightarrow (CH_3)_2CH{-}O{-}C(CH_3)_2OOH \longrightarrow\longrightarrow$$

O2, O2, O2

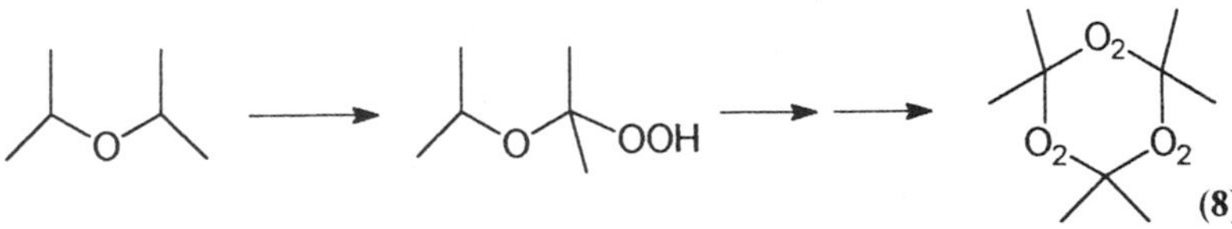

(**8**)

[1]It will be seen that this is very similar to the representation of the transition state for peroxyester decomposition discussed in Chapter 3, p. 28.

One very clear manifestation of polar effects in hydrogen-atom transfers is found in the rates of reactions of *para*-substituted toluenes with oxygen-centred radicals such as t-butoxyl. Very good correlations have been found with substituent constants (σ^+) which are indicative of conjugative interaction of the substituents with an electron-deficient centre. The ρ-value in this system was ca. −0.4. Some data are plotted in Fig. **4.10**, and the effect is illustrated by means of the transition state for abstraction from *p*-phenoxytoluene (**9**).

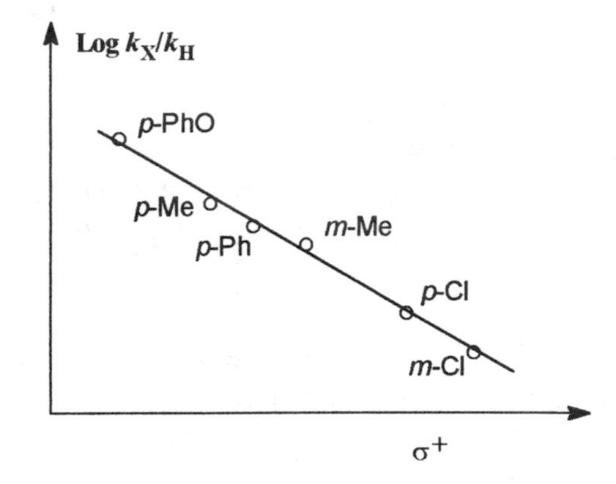

Fig. 4.10: Selected rate data for hydrogen abstraction by t-butoxyl radicals from the methyl group of ring-substituted toluenes plotted against the σ^+ substituent constants.

(9)

That oxygen-centred radicals are generally electrophilic is unremarkable. That simple alkyl radicals are somewhat nucleophilic may appear rather more surprising. An example of this is found in the free-radical alkylation of protonated pyridines and quinolines, in which reaction occurs exclusively at the α- and γ-positions – sites of particularly low π-electron density in the protonated heterocycles, e.g. Scheme 4.2.

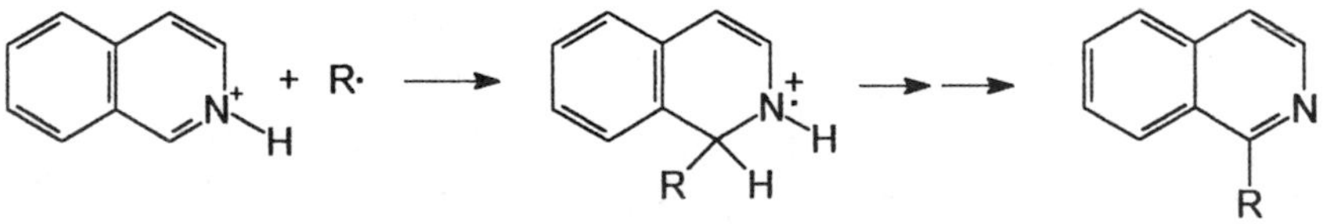

Scheme 4.2

Sole product from protonated isoquinoline

1 cal = 4.184 J

Of course, halogen atoms are electrophilic, so that the very low activation barrier for hydrogen abstraction from methane by chlorine atoms, discussed earlier in this chapter, is in part due to a transition-state contribution from $[Cl^- \; H\cdot \; CH_3^+]$. The barrier for the identity reaction: $CH_3\cdot + H\text{-}CH_3 \rightarrow CH_3\text{-}H + CH_3\cdot$, for which the transition state must be symmetrical, is very much higher (ca. 14 kcal mol^{-1}). This is important. Were it otherwise, many reactions in which alkyl radicals are generated in the presence of alkanes would be expected to exhibit extensive intermolecular hydrogen scrambling.

It has frequently been written that polar effects are important only when the transition state is early. There seems little rational basis for this conclusion, which is certainly inconsistent with a comparison between the above identity reaction and the slightly endothermic reaction of methane with chlorine atoms.

Intermolecular, as well as intramolecular, additions to double bonds, especially those of alkyl radicals, have become important in synthesis in recent years (Chapters 6 and 7). In the majority of instances the reactions are quite exothermic, but, as we saw in the case of methyl addition to ethene, there is a significant activation barrier. This implies considerable scope for rate manipulation by polar effects. Thus it is found that the additions of nucleophilic alkyl radicals to an alkene are greatly facilitated by the presence of an electron-withdrawing, e.g. cyano or alkoxycarbonyl, substituent in the alkene at the site β to the point of attack.[1] These effects, as well as steric effects (see below), are so pronounced in radical additions that they usually outweigh any simple thermochemical analysis of the kind presented for halogenation.

A beautiful example of polar effects in radical addition to alkenes is a reaction involving the hydroperoxide (**10**), α-methylstyrene, acrylonitrile and methanol (as solvent). All four reactants were shown to be incorporated into a single product (**11**) in good yield (ca. 60%). This reaction, which requires an iron salt as catalyst, illustrates in a quite remarkable way not only polar effects, but also many of the reaction types which we have encountered in earlier chapters. The reaction sequence is outlined in Scheme 4.3 (overleaf). Peroxide decomposition is initiated by means of Fe(II); the resulting alkoxyl radical undergoes a ring-opening fragmentation to give an alkyl radical. As we have just seen, alkyl radicals behave as nucleophiles, so that preferential addition to the acrylonitrile is unsurprising This gives a cyanoalkyl radical which, by comparison, is rather electrophilic and seeks out the α-methylstyrene; the result of this second addition is a tertiary benzylic radical. This is the first species in the sequence with an oxidation potential

[1]This has usually been believed to hold for all simple alkyl radicals, although consideration of cation stability would lead one to expect that the most nucleophilic alkyl radicals would be tertiary. Results of recent high-level MO calculations show rather clearly that addition of *methyl* radicals to alkenes is essentially devoid of any polar effect. Calculated activation barriers correlate well with reaction exothermicities for addition to alkenes ranging from vinylamine to acrylonitrile.

sufficiently low that it can be oxidised by the Fe(III) formed in the peroxide decomposition step – and regenerate Fe(II).

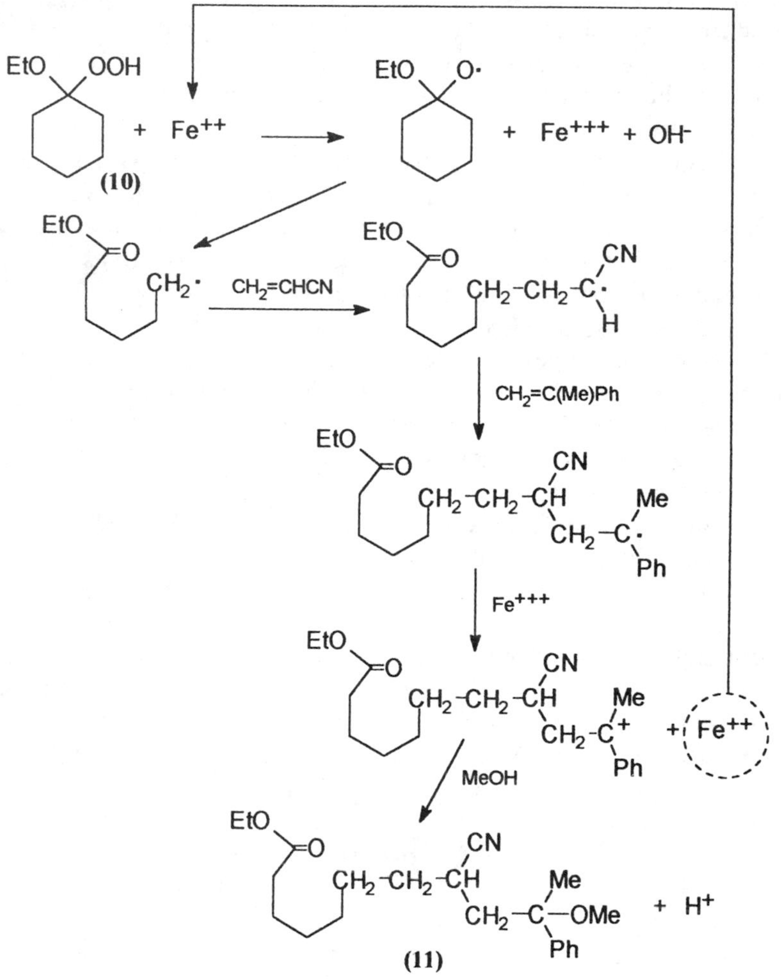

Scheme 4.3

We shall see in Chapter 9 that, because addition reactions of alkyl radicals generally involve an exothermic initial step (with an early transition state – a relatively weak C–C π-bond is replaced by a much stronger C–C σ-bond), an alternative interpretation of polar effects in these reactions is possible based on frontier orbital theory.

1 cal = 4.184 J

4.6 STERIC EFFECTS

Not surprisingly, steric effects can influence the ease of free radical transformations. The tri-isopropylmethyl radical (**12**), whilst superficially a simple tertiary alkyl radical, is in fact so congested round its radical centre that any reaction with itself is much slower than the diffusion-controlled limit (only ca. 10^3 mol^{-1} sec^{-1} at 300 K). A special term has been adopted to describe radicals whose self-reaction is significantly less than the diffusion limit. They are said to be "persistent". This term should not be confused with "stabilised". The benzyl radical (Fig. **4.1**) enjoys a significant measure of resonance stabilisation, but its dimerisation is essentially diffusion-controlled. It is not a persistent species. Radical (**12**) is not inherently

$(Me_2CH)_3C\cdot \equiv$

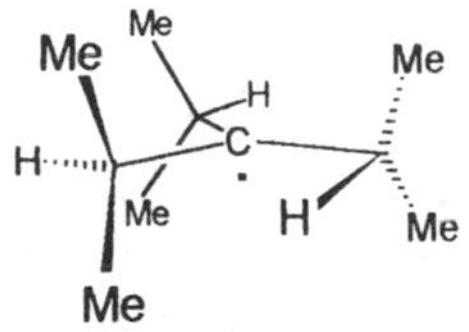

(12)

stabilised with respect to other tertiary alkyl radicals (although tri-isopropylmethane must experience a steric *de*stabilisation), but is relatively persistent. The persistence of radical species is reflected in the relatively high concentrations which may be maintained in solution. This usually makes spectroscopic study straightforward (see Chapter 5), and can dramatically influence product distribution from reactions in which such radicals are intermediates (see Section 9.8).[1]

In radical additions to alkenes, a pronounced steric effect arises from substituents at the point of attack. Thus addition of a primary alkyl radical to 2-methylbutene is very much slower (by a factor estimated to be ca. 50 at 300 K) than to 2-methylpropene. As noted earlier, both steric *and* polar effects play important roles in determining the actual outcome of radical additions to alkenes.

Less dramatically, in the reaction of phenyl radicals with t-butylbenzene (see Chapter 1), the proportion of *ortho* isomer in the resulting mixture of t-butyl-

Table 4.4: Proportions of isomeric biphenyls formed on free radical phenylation of toluene and t-butylbenzene.

Substrate	% *o-*	% *m-*	% *p-*
toluene	63	21	16
t-butylbenzene	24	49	27

[1]It should also be pointed out that the term "unstable" is an inappropriate descriptor for many short-lived radicals. Unstable is more usually used in the sense of unimolecular instability. In the vacuum of outer space a methyl radical would be very long-lived indeed!

biphenyls is significantly reduced in comparison with the results for phenylation of toluene; product isomer proportions for the two reactions are given in Table 4.4. The effect here is not particularly large, but it should be recognised that phenyl radicals are amongst the most reactive (and least selective) of carbon-centred radicals (from Table 4.3 it will be seen that the Ph–H bond is significantly stronger than the Me–H bond, a consequence of the changed carbon hybridisation). The great strength of aromatic C–H bonds also accounts for the fact that hydrogen abstraction from benzene is almost never observed. The same is true for vinylic, acetylenic, and also cyclopropyl[1] C–H bonds.

4.7 INTRA- VERSUS INTER-MOLECULAR REACTIONS

It was pointed out in Chapter 2 that many examples of radical rearrangements may be viewed as intramolecular radical additions or intramolecular atom transfers. The hexenyl → cyclopentylmethyl radical rearrangement is a case in point. It seems only natural that one should endeavour to compare not only the regiochemistry of inter- and intra-molecular processes, as we have done in the above consideration of stereoelectronic effects, but also the actual rates of the two types of reaction. Such a comparison is often assessed in terms of the "effective molarity" of the neighbouring group in the intramolecular reaction.

Effective molarity (sometimes "effective concentration") is defined as the ratio of the rate constant of the intramolecular reaction to the rate constant of a suitable intermolecular model (k_{intra}/k_{inter}). This has the dimensions of concentration.

Imagine a reactive species, such as a radical, present in low concentration, reacting bimolecularly with 1 M substrate. The disappearance of the radical will be pseudo-first order. Now suppose that the related intramolecular reaction, which is genuinely first order, proceeds at the same rate. It would then be said that the effective concentration of the neighbouring group in the intramolecular process is 1 M. Were the intramolecular process to have been ten times faster than the intermolecular one, the effective concentration would be 10 M, and so on.

In the case of hexenyl radical cyclisation, there is no suitable intermolecular model. This is because no reliable rate data appear to be available for the addition of a primary alkyl radical to carbon-2 of a terminal alkene. However, a comparison of published rate data is possible for the two reactions shown opposite. The result is an effective concentration in the intramolecular reaction of ca. 10^5 M. Such large numbers are well documented for ionic reactions, and their magnitude arises from entropic effects associated with reduced degrees of freedom for the reacting components of the intramolecular process. The chelate effect in inorganic chemistry has a similar origin.

[1]It should be remembered that in cyclopropane the hybrid orbital on carbon which is involved in bonding to hydrogen approximates to sp^2.

1 cal = 4.184 J

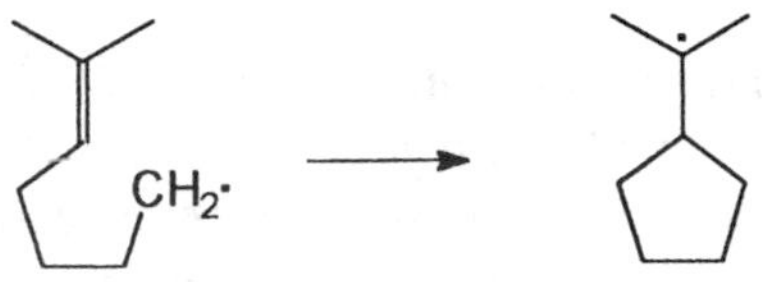

Intramolecular

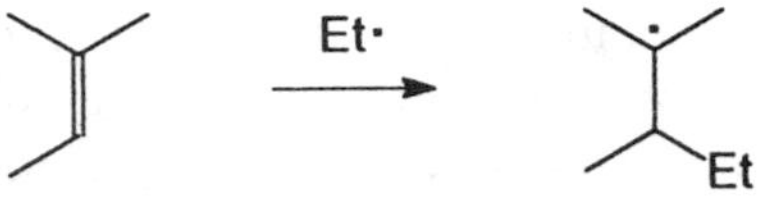

Intermolecular

The occurrence of large effective concentrations in intramolecular radical additions has been highlighted here in part because there are some reactions of this kind which have no intermolecular counterpart: it is only the kinetic advantage of intramolecularity that makes these reactions observable. An example is the intramolecular addition of a primary alkyl radical to the carbon of an aldehyde group, particularly when the ring formed is five- or six-membered, e.g. Scheme 4.4. In this example, the success of the cyclisation reflects the very much higher rate of reaction with tributyltin hydride[1] of the electrophilic alkoxyl radical than of the precursor alkyl radical. This enables the trapping of the alkoxyl radical to compete successfully with its reversion to alkyl radical – which is actually much faster than the cyclisation.

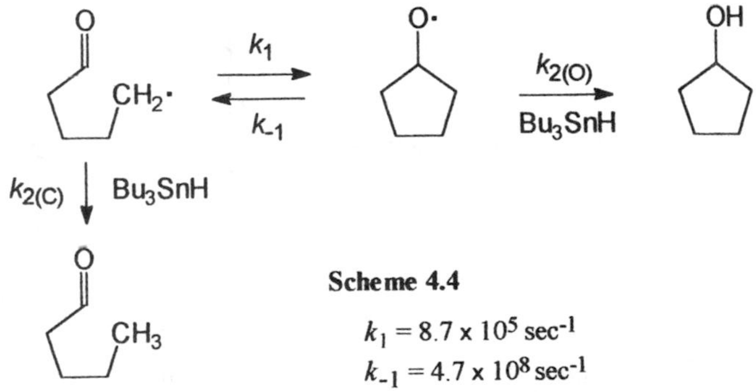

Scheme 4.4

$k_1 = 8.7 \times 10^5\ \text{sec}^{-1}$

$k_{-1} = 4.7 \times 10^8\ \text{sec}^{-1}$

$k_{2(O)} = 3.7 \times 10^8\ \text{M}^{-1}\ \text{sec}^{-1}$

$k_{2(C)} = 3.2 \times 10^5\ \text{M}^{-1}\ \text{sec}^{-1}$

[1]The chemistry of this important reducing agent is discussed further in Chapter 6.

4.8 RADICAL ADDITION VERSUS HYDROGEN ATOM TRANSFER

Radical attack on alkenes can take either of two pathways, namely addition to the π-system or abstraction of an allylic hydrogen. Reversibility is often important here too, a point which is nicely illustrated by further consideration of allylic bromination by *N*-bromosuccinimide (NBS) which was mentioned in Chapter 1. The success of this reaction depends upon the ability of the reagent to provide a souce of a constant *low concentration* of molecular bromine. Evidence for this comes in part from the successful achievement of allylic bromination by direct introduction of low concentrations of Br_2 instead of by using NBS. With higher bromine concentrations the predominant result is addition. The interpretation of this behaviour should be evident from an examination of Scheme 4.5. This shows that the addition of Br· is faster than abstraction of the relatively weakly held (Table 4.3) allylic hydrogen, but also that it is rapidly reversible. The consequence is that products of addition will be important only if the initial β-bromoalkyl radical (**13**) is rapidly scavenged by high concentrations of bromine. When the concentration of bromine is very low, product formation is by the slower abstraction route – which is effectively irreversible. In the NBS system the HBr produced reacts with NBS forming succinimide and maintaining the low concentration of bromine.

HBr + Br$_2$ Br

Br· +

Br$_2$ Br Br

Br

(**13**)

O O

N—Br + HBr ⟶ N—H + Br$_2$

O O

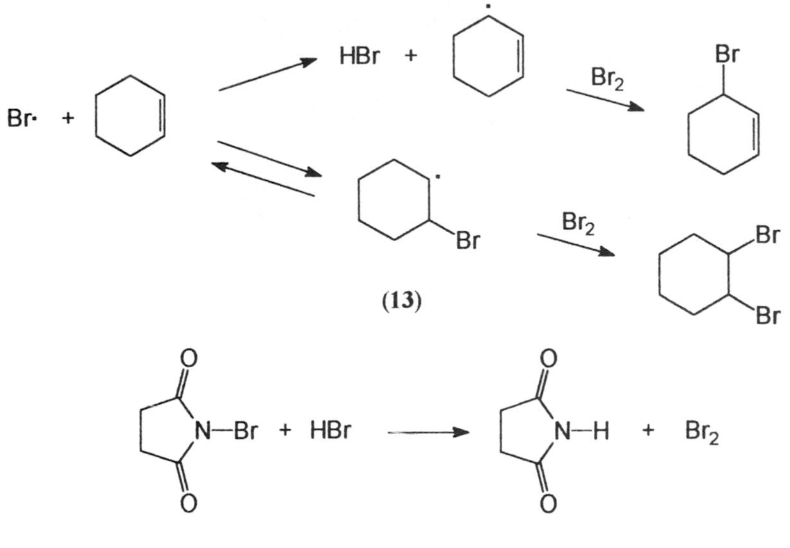

Scheme 4.5

A more subtle discrimination comes into play in cases in which both processes are effectively irreversible. As yet, there seems to be no qualitative picture which can rationalise all of the available data, although some progress has been made in understanding variations exhibited by closely related radicals.

1 cal = 4.184 J

Benzylic abstraction and addition to the benzene ring may compete in a similar manner in reactions of toluene and other aralkanes, although here it is generally only the most reactive radicals (e.g. HO·, Ph·, and Me·) which are effective in the addition process.

There remains one final consideration which should not be overlooked when relative reactivity data are being discussed. Very often the rates being compared are rather similar, and closer scrutiny may reveal that the observed differences are dominated by entropy effects rather than enthalpy effects. This will sometimes lead to an inversion in reactivity orders when the temperature of comparison is varied. The point at which the reactions have identical rates is called the "isoselective temperature".

4.9 SUMMARY

Whilst, to a first approximation, it may be possible to relate the ease of a simple radical process (e.g. atom transfer, or addition to a π-system) to the difference between the strength of bond being broken and the strength of bond being formed, other kinetic effects are very important. These include polar, steric, and stereoelectronic factors.

5

Detection of organic free radicals, and a discussion of their structures

5.1 INTRODUCTION

The unpaired electron imparts paramagnetic properties to radicals. However, only with stable radicals, where high concentrations are possible, can these properties be detected by means of simple magnetic susceptibility measurements. Where it is suspected that reactive radicals may be present as low-concentration transients in a chemical reaction, or perhaps in a biological system, some more sensitive procedure is necessary.

In early chemical investigations, radical participation in a reaction was sometimes inferred if the reaction could be catalysed by a typical initiator such as a peroxide or an azo-compound, or from the ability of the reaction system to promote polymerisation of a suitable additive, such as styrene. Another useful technique involved monitoring the decolorisation of a stable radical, such as the purple diphenyl picryl hydrazyl (see p. 26) or the brown phenoxyl (**1**), known as "galvinoxyl",[1] which had been added to the reaction as a radical scavenger. In

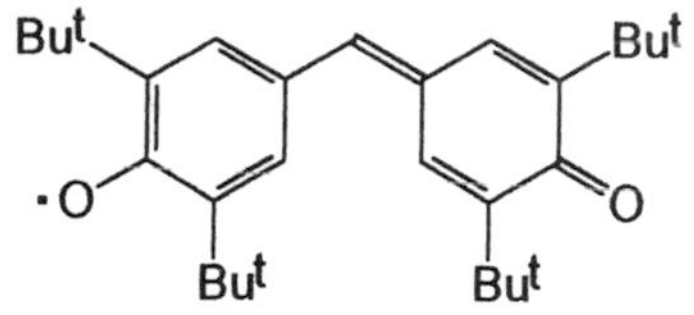

(1)

favourable circumstances, monitoring the disappearance of this colour has afforded a means of determining the rate of decomposition of a radical initiator. On the other hand, for some radical processes which have very high rate constants for individual

[1]After its discoverer, Galvin Coppinger.

1 cal = 4.184 J

steps, the polymerisation approach is doomed to failure (see, for example, the addition of CBr_4 to styrene described in Chapter 2), and even the stable radical technique may fail.

Our principal concern in this chapter will be with two magnetic resonance spectroscopies. The student of NMR quickly learns that paramagnetic impurities in a sample lead to line broadening and a loss of resolution. For optimum resolution, removal of dissolved oxygen may even be necessary. Also, NMR is a relatively low-sensitivity spectroscopy. Therefore, it may come as some surprise that NMR methodology can give important information about radical reactions – using a phenomenon known as CIDNP (pronounced "sidnip" or "kidnap"). But before explaining CIDNP, we shall examine ESR, or electron spin resonance,[1] which looks directly at the magnetic properties of the unpaired electron in a radical species. In doing this we shall make comparisons with conventional NMR spectroscopy and shall assume some familiarity with that technique.

5.2 ESR SPECTROSCOPY

Whereas NMR depends on reorientation of magnetic *nuclei* in the presence of a magnetic field, ESR depends on reorientation of unpaired *electrons* in a magnetic field. Like many atomic nuclei, unpaired electrons have "spin". Associated with a spinning electrically charged particle is a magnetic moment. The spin quantum number of a single electron is ½, so that only two orientations of the "electron magnet" are possible in the presence of an applied magnetic field. These two magnetic "spin states" differ in energy. Detection of the presence of unpaired electrons is then possible by monitoring the absorption of electromagnetic radiation the energy of which corresponds to the energy difference between the two spin states. Commonly, electron spin resonance spectrometers operate with magnetic field strengths of ca. 3300 gauss (0.33 tesla). This is much lower than the fields used in NMR spectrometers, but the magnetic moment of the electron (μ_B) is very much greater than the magnetic moments of atomic nuclei, and the electron resonance frequency at this field strength is ca. 9000 MHz (9 GHz).[2]

Just as nuclear resonance signals may show hyperfine splitting as a result of interaction of the observed nucleus with near-neighbour nuclei which are also magnetic, so electron resonance signals may show hyperfine splittings for a similar reason. For example, just as the aldehyde proton in acetaldehyde exhibits a quartet signal (Fig. **5.1a**) because of interaction with three equivalent methyl protons, so the electron resonance signal of a methyl radical is correspondingly split into a quartet (Fig. **5.1b**). The spacing between the lines, termed the hyperfine splitting constant

[1]Also known as electron paramagnetic resonance (EPR).

[2]This frequency is in the X-band region of microwave radiation, the technology of which was originally investigated for the development of radar during the Second World War. The corresponding wavelength is 3 cm. Higher magnetic fields and correspondingly higher frequencies have also been used in commercial spectrometers.

(a), is commonly recorded in magnetic field units (gauss or millitesla). For methyl radicals, a_H (the subscript indicates splitting by protons) is ca. 23 gauss (2.3 mT).

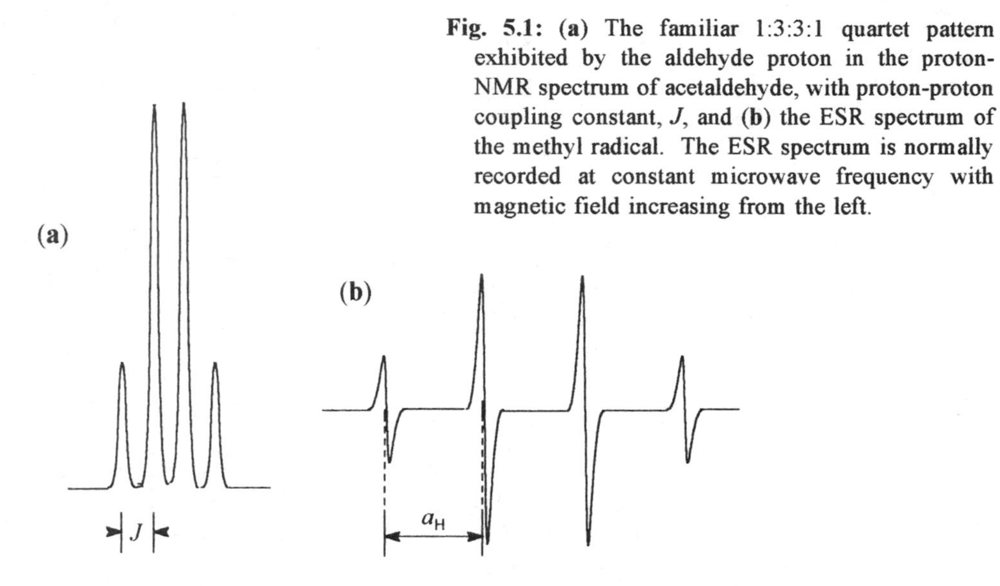

Fig. 5.1: (**a**) The familiar 1:3:3:1 quartet pattern exhibited by the aldehyde proton in the proton-NMR spectrum of acetaldehyde, with proton-proton coupling constant, J, and (**b**) the ESR spectrum of the methyl radical. The ESR spectrum is normally recorded at constant microwave frequency with magnetic field increasing from the left.

One immediately obvious difference from NMR is the appearance of the ESR signal. This is presented as the first derivative of the absorption signal; i.e. what is plotted against magnetic field is the gradient of the absorption signal. As the magnetic field is changed, this gradient peaks (A in Fig. **5.2**) well before maximum absorption, by which time it has fallen back to zero (B in Fig. **5.2**); it then peaks in the negative direction (C) before returning to the base line. Historically, this format was for reasons of instrument design, based on the requirements of signal detection circuitry.[1] Although this is no longer necessary the format has been retained, not least because it has the effect of increasing resolution between two overlapping peaks (Fig. **5.3a** and **b**). For the same reason it is not unusual to see second derivative ESR spectra recorded (inverted), where the effect is even more striking (Fig. **5.3c**).

[1]Microwave radiation is fed to a resonant "cavity" which contains the sample cell and is located in the magnetic field. The dimensions of the cavity are such that a standing wave pattern is established. As the magnetic field is varied, and the resonance condition is traversed, the standing wave in the cavity interacts with the paramagnetic sample and there is a small reduction in the amount of microwave energy which is reflected from the cavity to a microwave detector. The method by which this small diminution in signal is measured results in the first derivative trace that has been described.

1 cal = 4.184 J

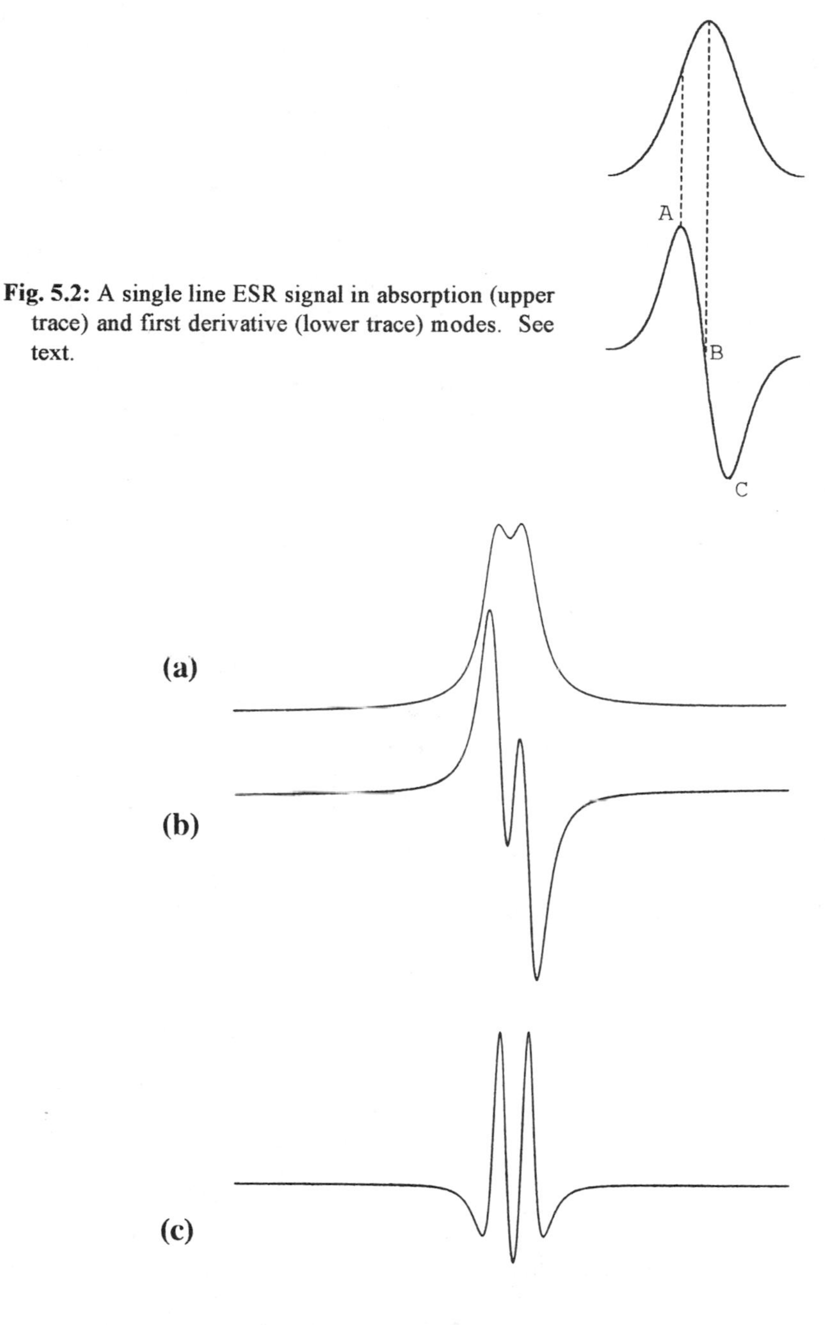

Fig. 5.2: A single line ESR signal in absorption (upper trace) and first derivative (lower trace) modes. See text.

Fig 5.3: Computer simulation of two overlapping peaks of equal intensity shown (**a**) in absorption; (**b**) as the first derivative of the absorption, and (**c**) as the second derivative (with the y-axis inverted). Identical line spacings and line-shape and line-width parameters were used for each calculation.

Two other comparisons with NMR are instructive. The techniques are similar in that symmetrical, highly resolved spectra are obtained only when the radical containing the unpaired electron is tumbling rapidly in solution. This has the effect of averaging out magnetic anisotropy effects.[1] Although there is a very considerable literature on the ESR spectra of paramagnetic materials in the solid state, this is almost entirely beyond the scope of the present survey. We shall also pay little attention to the ESR analogue of the chemical shift. This is the experimentally determined g-factor (sometimes g-value) of the radical in Equation (1), which relates resonance frequency, ν, to the applied magnetic field, B. The magnetic moment of the electron, μ_B, is a constant referred to as the "Bohr magneton". For rapidly tumbling organic radicals, g-values are always close to 2.0 and are only slightly dependent on radical structure, so that if two or more radicals occur in detectable concentrations in a single sample, their spectra will be seen to overlap.

$$h\nu = g\mu_B B \quad (1)$$

The ESR spectrum of the relatively simple triphenylmethyl radical actually appears very complicated. This is because the unpaired electron interacts with all 15 protons. The 6 *ortho*-hydrogens will give a septet, as will the 6 *meta*-hydrogens – but with a different hyperfine splitting constant; finally, the 3 *para*-hydrogens will give a quartet. The result is a "quartet of septets of septets", i.e. $4 \times 7 \times 7$ or 196 lines! Not all of these are resolved (Fig. **5.4**); many of the smaller ones which occur near the centre of the spectrum are swamped by more intense lines which occur in that region.

A simpler example is given in Fig. **5.5**, which shows the spectrum of the naphthalene radical anion. In this case the electron interacts with two non-equivalent sets of four protons each, resulting in a quintet of quintets. All of the expected 25 lines can be seen in this spectrum. Reviewing the binomial distribution of intensities of lines in a multiplet, learnt from analysis of NMR spectra, the reader should recognise the 1:4:6:4:1 patterns in each of the quintets.

Whilst it is a relatively straightforward matter to record spectra of persistent radicals like $Ph_3C\cdot$,[2,3] any transients which, in solution, decay by self-reaction at the

[1]Spectral line-widths are greater than in NMR, so that magnet design is less critical, and there is no requirement for sample spinning.

[2]The significance of the descriptor "persistent" was explained in Chapter 4, p. 47.

[3]Oxygen must be excluded from the sample, both because it may react with the radical, and because electron exchange may cause line broadening. Qualitatively, pairs of electrons are switched between radicals *via* the relatively high concentration of oxygen. If this phenomenon is rapid, the electrons lose information about the spins of nuclei with which they have been associated. The same "exchange broadening" may occur in the absence of oxygen if the radical concentration is too high (ca. 10^{-4} M, depending on the natural line-width of the spectrum).

diffusion-controlled limit are unlikely to reach concentrations greater than ca. 10^{-8} M (see Section 2.4), which is just below the normal detection limit of the spectrometer. This usually lies in the range 10^{-7}–10^{-6} M. Since isotropic ESR spectra give useful information on the structure of free radicals, extensive efforts were made in the 1960s to develop methods whereby concentrations could be raised above the

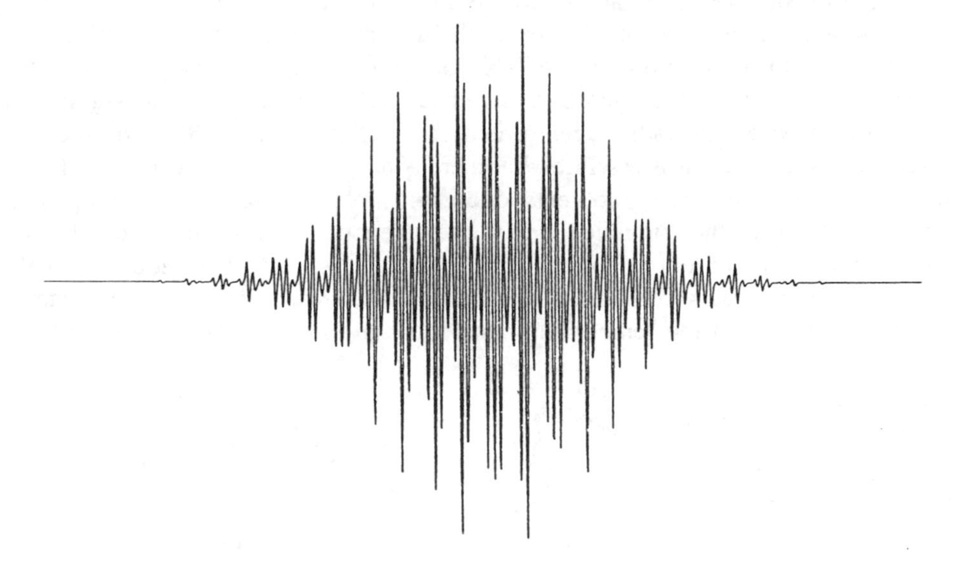

Fig. 5.4: The ESR spectrum of triphenylmethyl.

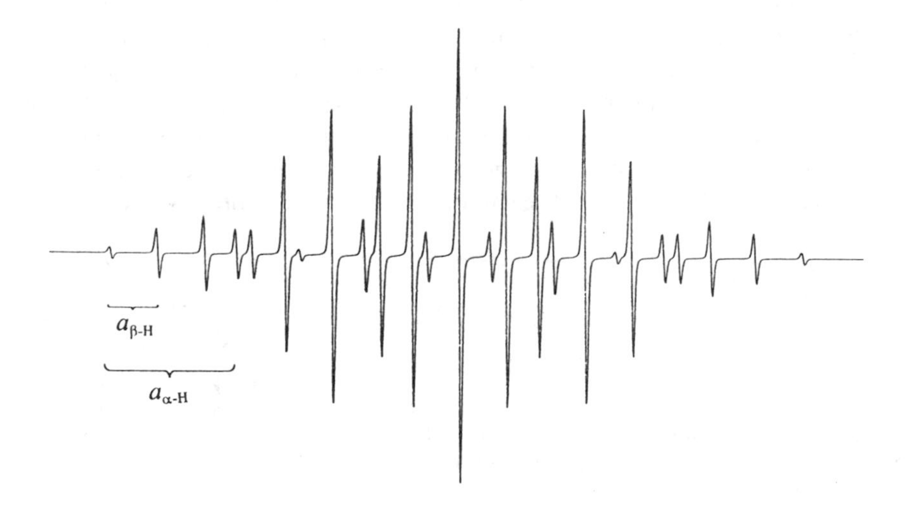

Fig. 5.5: The ESR spectrum of naphthalene radical anion, clearly showing the small 1:4:6:4:1 quintet splitting from the β-protons, and the larger quintet splitting from the α-protons ($a_{\alpha\text{-H}}$ = 4.95; $a_{\beta\text{-H}}$ = 1.85 gauss).

detection limits of the equipment. One of these, which has also been adapted for direct measurement of kinetics of unit steps in radical processes, involves the simple expedient of focusing the output of a powerful ultraviolet lamp onto the sample whilst it is resident in the spectrometer. Under these conditions, photodissociation of a suitable radical source may be sufficiently rapid that the stationary state concentration of a reactive radical will build to levels quite adequate for good-quality spectra to be recorded. For example, decomposition of the peroxyester **(2)** in this way yielded a spectrum of the 7-norbornenyl radical **(3)**. Comparison of this spectrum with that of the 7-norbornyl radical failed to indicate any significant interaction between the radical centre and the double bond in **(3)**. For such an experiment to succeed, it is essential that there is no rapid reaction with the solvent. Suitable, relatively unreactive solvents for organic molecules have included liquefied cyclopropane and the liquefied inert gas, xenon. These experiments are conventionally carried out at sub-ambient temperatures, in part to reduce radical–molecule reaction rates. Also, the population difference between the spin states increases with decreasing temperature, and this can give a detectable increase in sensitivity.

$$\xrightarrow[-CO_2]{h\nu}$$

(2) **(3)**

Although water is unreactive towards organic radicals, it is also a poor solvent for many organic radical precursors. For ESR experiments, there is additionally the technical problem of microwave absorption by the water. There are several experimental techniques for overcoming this, one or other of which is usually essential in biological investigations. They are also incorporated into a second procedure for "pumping" the rapid production of radicals in which a "flow-through" sample cell is used. Aqueous solutions of rapidly reacting precursors are mixed immediately prior to their entering the cell, which is already positioned in the spectrometer (Fig. **5.6**). The rapid reaction, usually of the Fenton type (Chapter 3),

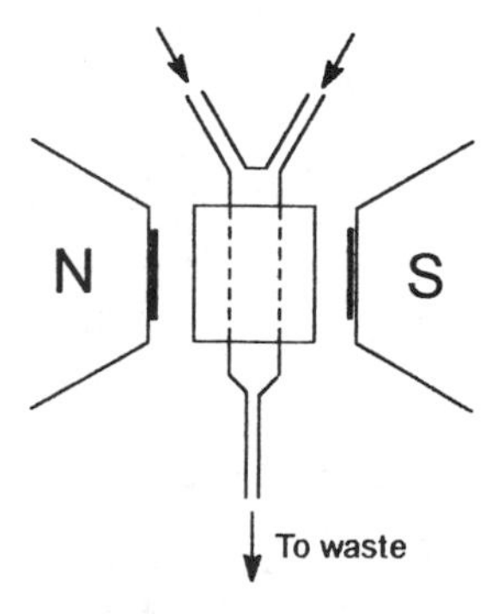

Fig. 5.6: Schematic representation of mixing cell for an ESR spectrometer. The cell is partially surrounded by the microwave resonant cavity.

1 cal = 4.184 J

generates hydroxyl radicals which themselves react with the organic substrate of interest. Radical formation and decay is substantially complete before the mixture exits the cell, but concentrations within the cell are again sufficiently high for radical detection to be relatively straightforward. The solutions must, of course, be kept flowing whilst spectra are recorded, so that rather large reaction volumes are usually necessary. An example, in which Fe(II) is replaced by Ti(III), is given in Scheme 5.1 and Fig. **5.7**.

$$Ti^{3+} + H_2O_2 \longrightarrow Ti^{4+} + HO\cdot + HO^-$$

$$HO\cdot + CH_3CH_2OH \longrightarrow CH_3\dot{C}HOH + H_2O$$

Scheme 5.1

Both the photolytic and the flow-through procedures are quite inexpensive when compared with the first successful attempt to record ESR spectra of transient radicals in solution. In that pioneering work, the high concentrations of transients were maintained by bombardment of the sample with an intense beam of electrons from a powerful Van de Graaff generator.

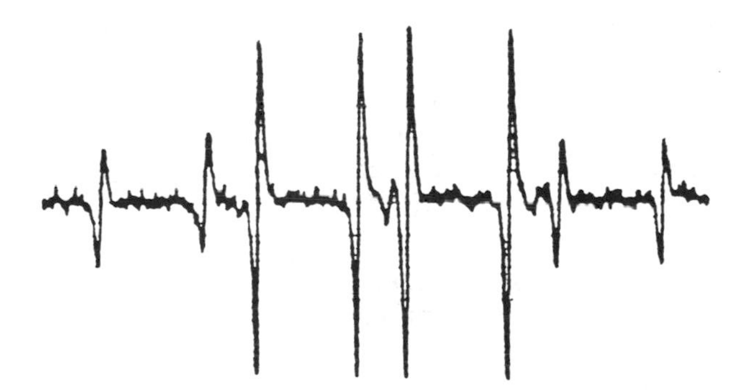

Fig. 5.7: The ESR spectrum of 1-hydroxyethyl radicals from the reaction of hydroxyl radicals with ethanol using the flow-system technique. Note that there is no observable hyperfine splitting from the hydroxyl hydrogen. (Reproduced with permission from W.T. Dixon and R.O.C. Norman, *J. Chem. Soc.*, **1963**, 3121).

5.3 ESR AND RADICAL STRUCTURE

From the spectrum of the naphthalene radical anion (Fig. **5.5**) it might reasonably be inferred that the unpaired electron is more strongly associated with the α-position than with the β-position. This is entirely consistent with molecular orbital calculations which place the odd electron in a π^*-antibonding orbital having its largest p-orbital coefficients at C-α. The α-position is said to have the highest π-electron spin population.[1] For planar aromatic systems like this, there is an empirical relationship between a_{H} and π-electron spin population, ρ_{π}, at the adjacent carbon atom [Equation (2)].

$$a_{\mathrm{H}} = Q\rho_{\pi} \tag{2}$$

Based on the methyl radical, the planarity of which has been established from the rotational fine structure in its gas-phase electronic spectrum,[2] and in which all of the spin population is therefore in the carbon p-orbital, Q should have a value of 23 gauss. A value for Q based on the spectrum of the naphthalene radical anion ($a_{\alpha\text{-H}}$ = 4.95 gauss; $a_{\beta\text{-H}}$ = 1.85 gauss) and its theoretically determined unpaired-electron distribution, would be some 15% greater than this; despite this discrepancy the correlation has proved to be a useful one.

It might reasonably be asked why there should be *any* proton splitting in these radicals at all: this is because coupling is transmitted by way of the bonding electrons, but the unpaired electron is located in an orbital *orthogonal* to the sp^2-orbitals whereby the hydrogen atoms are attached! The answer is pictured in terms of what is called "spin polarisation". This is illustrated in Fig. **5.8**. The pair of electrons in the C–H bond are spin polarised in such a way that the one having its spin parallel to that of the unpaired electron spends more time at the carbon end of the bond (a manifestation of Hund's Rule); the electron which spends more time associated with hydrogen is therefore of opposite spin to the odd electron, so that the hyperfine coupling to hydrogen is "negative".

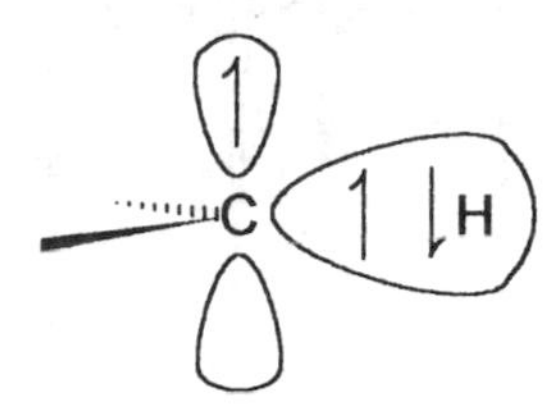

Fig. 5.8: "Spin polarisation" in the C–H bond of a π-radical.

[1]Sometimes, but incorrectly here, "spin density".

[2]Chemical evidence presented in support of the planarity of alkyl radicals derives *inter alia* from the greater ease of forming t-butyl than bridgehead t-alkyl radicals. For example, at 60°C, the decomposition of (**i**) is 1000 times slower than that of (**ii**) (see the discussion of peroxyester decomposition in Chapter 3, and Problem 1). There is, however, compelling theoretical evidence that deformation from planarity is far easier than in the case of carbocations.

O O Bu-t (**i**)

t-Bu O O Bu-t (**ii**)

1 cal = 4.184 J

In the spectrum of the ethyl radical, the hyperfine splitting due to the β-hydrogens is actually larger than that due to the α-hydrogens attached to the radical centre. This β-hyperfine interaction is positive, and arises from electron delocalisation into the β C–H bond by hyperconjugation (Fig. **5.9**). This delocalisation of the unpaired electron in alkyl-substituted methyl radicals may be

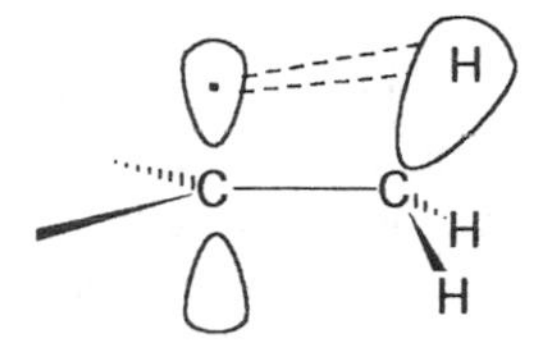

Fig. 5.9: Hyperconjugative interaction between an unpaired electron in a *p*-orbital on carbon, and the β C–H bond pair.

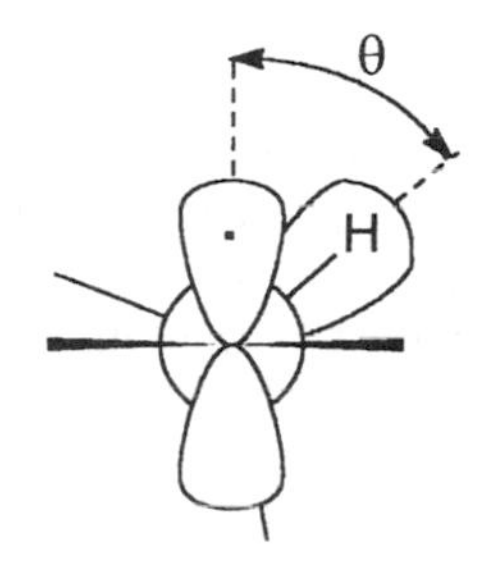

Fig. 5.10: The dihedral angle θ between the *p*-orbital on carbon and the β C–H bond.

associated with the increase of radical stabilisation in the sequence methyl < primary < secondary < tertiary.[1] The magnitude of the hyperconjugative interaction depends on the dihedral angle, θ, between the C–H bond and the axis of the *p*-orbital [Fig. **5.10**, and Equation (3) in which A and B are constants].[2]

$$a_{\beta\text{-H}} = \mathrm{A} + \mathrm{B}\cos^2\theta \tag{3}$$

[1]In Chapter 4 it was seen that the tertiary C–H bond, e.g. in 2-methylpropane, is weaker than primary or secondary C–H. Whilst hyperconjugation in the derived t-alkyl radical is undoubtedly a significant factor here, there will also be some relief of steric compression in going from tetrahedral 2-methylpropane to planar t-butyl radical which must also contribute to weakening the tertiary C–H bond.

[2]More generally, for any π-radical (see below), where not all of the spin population will be localised on the alkylated carbon atom, Equation (3) is replaced by (3').

$$a_{\beta\text{-H}} = (\mathrm{A} + \mathrm{B}\cos^2\theta)\rho_\pi \tag{3'}$$

Dihedral effects are averaged by rapid rotation of the alkyl group, and this conformational averaging is reflected in the magnitude of the β-hyperfine splitting. An extreme example is the triisopropylmethyl radical (p. 47), in which $a_{\beta\text{-H}}$ is only 2.4 gauss.

Radicals which have their unpaired electron located in a *p*-orbital, like methyl, or delocalised in a π-electron system, are referred to as π-radicals, but not all organic radicals are of this type. When the unpaired electron is in a hybrid orbital having some *s*-character, the species is commonly referred to as a σ-radical. This is the case, for example, in a trifluoromethyl radical, which, unlike methyl itself, is pyramidal, and where the electron is in a sp^3-orbital. Strongly electronegative substituents like fluorine or alkoxyl affect the geometry of an alkyl radical, as is revealed by changes in the hyperfine coupling to α-hydrogen or to ^{13}C substituted at the radical centre. As the *s*-character of the orbital containing the unpaired electron increases, this is reflected by increasingly positive values of the corresponding hyperfine splitting constants (Table 5.1).

For the last three entries in the table, the ^{13}C-splittings, whilst much larger than those expected of π-radicals, are also much smaller than would be expected if the unpaired electron were to be in an sp^2-orbital. This is least readily accommodated for the phenyl radical (**4**). In the other two examples the bond angle is substantially greater than 120°, so that the orbital containing the single electron does have increased *p*-character.

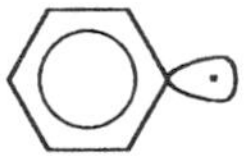

(4)

ESR spectroscopy has been combined with other physicochemical techniques, typically rotating-sector photochemical methods, to access directly rate constants for individual reaction steps. As we have noted elsewhere, many of these are very large, and, even today, relatively few absolute rate constants have been measured directly. Instead, much of the kinetic work described in the chemical literature has yielded only *relative* rate constants. For example, the relative reactivities of primary and tertiary hydrogens towards abstraction by a chlorine atom are readily determined by product studies (Chapter 4). Similarly, the rate of reaction of t-butoxyl radicals with toluene (by hydrogen abstraction from the methyl) can be compared with their rate of fragmentation [$\rightarrow CH_3\cdot + (CH_3)_2CO$] simply by comparing the yields of t-butanol and acetone (see Scheme 3.2). The accurate measurement of a number of absolute rate constants is therefore very important, since by combining the results with those of relative rate measurements a substantial catalogue of rate constants may be accumulated. There are, however, difficulties. Radical concentrations are not easy to measure, so that second-order rate constants for radical–radical reactions have been particularly difficult to access with precision. Much of the data in the

1 cal = 4.184 J

literature still depends on absolute measurements of questionable accuracy, possibly coupled with further errors which may have crept into the comparison experiments. Therefore errors of ± 100% may not be unusual, and sometimes even order of magnitude accuracy is doubtful. Amongst the most carefully studied reactions, and one which has been used extensively as a reference, is the first-order cyclisation of hexenyl to cyclopentylmethyl radicals. For this process the activation energy, E_a, is 6.85 kcal mol^{-1} and the pre-exponential factor in the Arrhenius equation is 2.6×10^{10} sec^{-1}; at 300 K these correspond to a rate constant of ca. 10^5 sec^{-1}.

Table 5.1: Isotropic hyperfine splittings (gauss) for radicals with general structure:

Radical	a_C	a_X (X=H)	a_{other}
$\cdot CH_3$	38.3	(–) 23.0	–
$\cdot C_2H_5$	39.1	(–) 22.4	26.9 ($a_{\beta\text{-H}}$)
$(CH_3)_3C\cdot$	45.2	–	22.7 ($a_{\beta\text{-H}}$)
$\cdot CH_2F$	54.8	(–) 21.1	64.3 (a_F)
$\cdot CHF_2$	148.8	(+) 22.2	84.2 (a_F)
$\cdot CF_3$	271.6	–	142.4 (a_F)
$(CH_3O)_3C\cdot$	152.7	–	0.4 (a_H)
$C_6H_5\cdot$	129	–	17.4 ($a_{o\text{-H}}$)
$CH_2{=}CH\cdot$	107.6	(+) 13.8	66 ($a_{trans\ \beta\text{-H}}$) 40 ($a_{cis\text{-}\beta\text{-H}}$)
$HC(=O)\cdot$	134.5	(+) 136.5	–

5.4 SPIN TRAPPING

An indirect use of ESR spectroscopy depends on the interception of reactive radicals (R·) by small quantities of *diamagnetic* molecules which have a very high affinity for radicals and react with them to give relatively persistent radical adducts. These diamagnetic radical scavengers are then known as "spin traps" (ST) and the resulting persistent radicals are referred to as "spin adducts" (R-ST·) [Equation (4)]. The spin adducts are normally sufficiently persistent to build up to concentrations which permit their detection (and, in favourable cases, unambiguous identification) by simple ESR methods. The most common of the spin traps in general use are nitrones and *C*-nitroso-compounds. In both cases the spin adducts belong to the class of radicals called nitroxides ($R_2NO\cdot$).

$$\underset{\text{Reactive radical}}{R\cdot} + \underset{\text{Spin trap}}{(ST)} \longrightarrow \underset{\text{Spin adduct}}{(R\text{-}ST\cdot)} \quad (4)$$

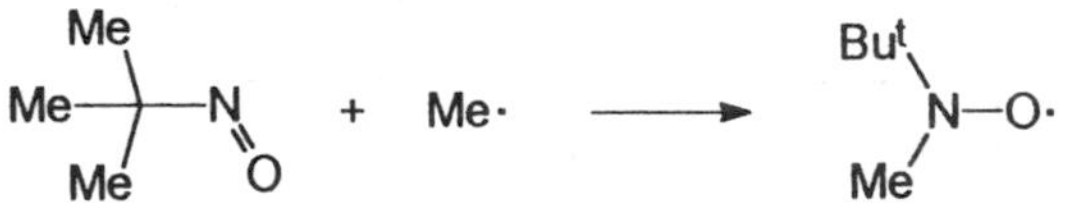

Me₃C–N=O (MNP) + Me· ⟶ Bu^t(Me)N–O·

(MNP)

(DMPO) + Me· ⟶ 2,2-dimethyl-5-methyl-5-H-pyrrolidine N–O·

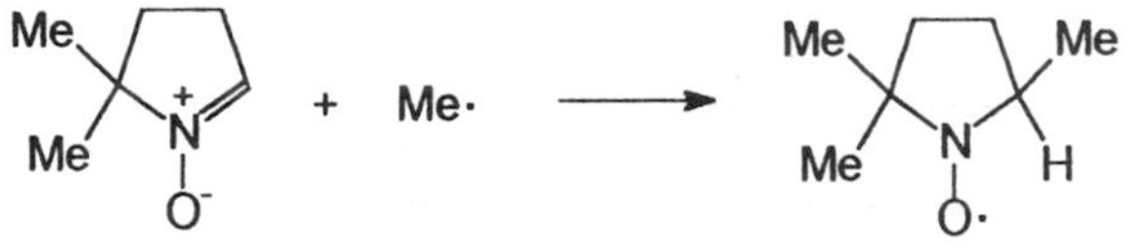

(DMPO)

Two examples are shown, the traps being 2-methyl-2-nitrosopropane ("MNP") and the cyclic nitrone, 5,5-dimethylpyrroline *N*-oxide ("DMPO") respectively. In each case the spectra of their spin adducts with methyl radicals are illustrated in Fig. **5.11**. Inspection of these spectra reveals, also in each case, something which we have not previously encountered, and which is uncommon in NMR spectroscopy. This is the clear 1:1:1 threefold multiplicity which arises from interaction of the unpaired electron with the nitrogen-14 nucleus. ^{14}N has nuclear spin, I, = 1, and, like other magnetic nuclei, can adopt $2I + 1$ different orientations in the magnetic field. In the first spectrum (Fig. **5.11a**) there is superimposed on the 1:1:1 triplet a clear 1:3:3:1 quartet pattern arising from the three equivalent methyl protons; in the second case this direct evidence for the presence of the methyl group is absent. The methyl protons in the DMPO adduct are more remote from the unpaired electron, and any small quartet splitting is hidden within the width of the spectral lines. What *is* apparent, in addition to the 1:1:1 triplet, is a doublet splitting, due to the pyrrolidine proton at the point of attachment of the methyl group (Fig. **5.11b**). The obvious advantage of the nitroso-compound in terms of the structural

1 cal = 4.184 J

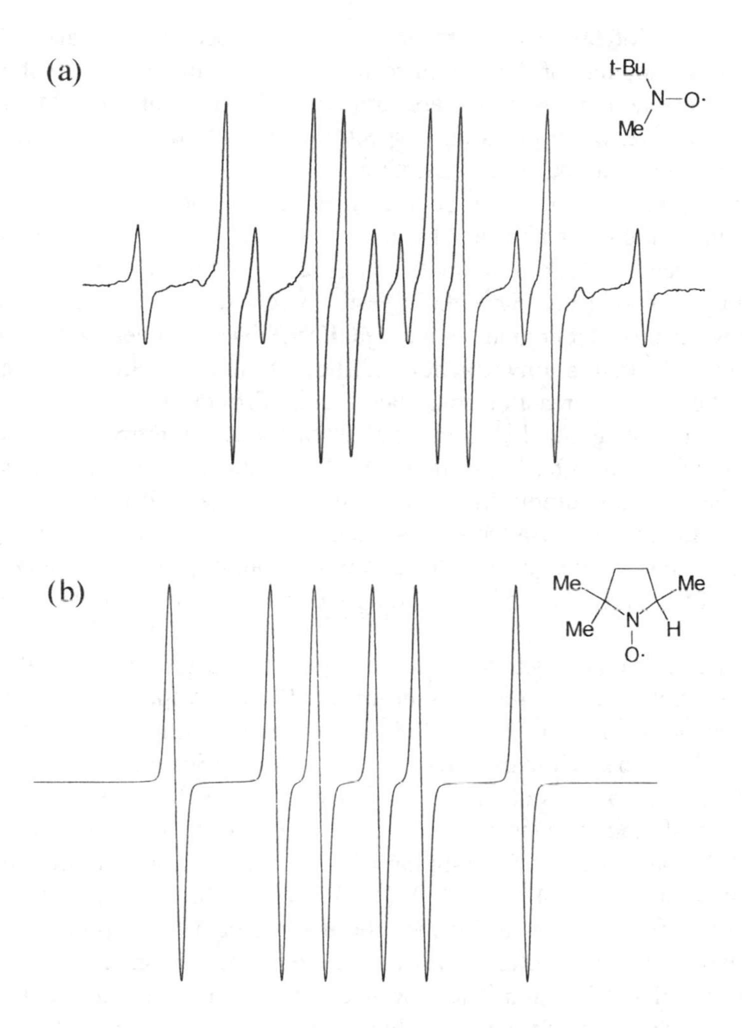

Fig. 5.11: The ESR spectra of methyl radical adducts of (**a**) MNP (a_N = 15.25; a_H = 11.3 gauss) and (**b**) DMPO (a_N = 14.3; a_H = 20.5 gauss).

information which may be deduced from the spectrum is offset by several major disadvantages. In the first place, MNP is photolysed by red light, so that it may be difficult to obtain spectra free from contamination by that of di-t-butyl nitroxide (see Scheme 5.2). More serious, however, is that in biological investigations (see

$$\mathrm{Bu^t{-}N{=}O} \xrightarrow{h\nu} \mathrm{Bu^t\bullet} + \mathrm{NO}$$

$$\mathrm{Bu^t\bullet} + \mathrm{Bu^t{-}N{=}O} \longrightarrow (\mathrm{Bu^t})_2\mathrm{N{-}O\bullet}$$

Scheme 5.2 — di-t-butyl nitroxide

Chapter 10) it is often desirable to monitor oxygen-centred radicals (HO·, HOO·, ROO·), but the adducts of these to nitroso-compounds are usually so short-lived as to be undetectable, if indeed they are formed at all. In contrast, DMPO, which is very water soluble, often gives good spectra of adducts to these species, although even those adducts are relatively unstable.

Unfortunately, detection of spin adducts does not constitute unambiguous evidence for a radical pathway. In the first place, both nitroso-compounds and nitrones are readily attacked by nucleophiles, and the resultant hydroxylamines (or their conjugate bases) are particularly easily oxidised to nitroxides. Furthermore, as with any sensitive technique, including CIDNP discussed below, a positive result may mean only that a tiny fraction of the reaction is following the observed pathway; the principal reaction route being quite different!

In view of the generally rather stable character of nitroxide radicals, spin-trapping experiments often give mixtures of spin adducts whose ESR spectra are superimposed. In favourable cases, where the adducts are all reasonably persistent, it has proved possible to use the spectrometer as an hplc detector, stopping the flow each time a paramagnetic species elutes from the column, and recording its spectrum uncontaminated by spectra of other components of the original mixture (e.g. Fig. **5.12**).

A final example shows how it has been possible to estimate the rate of a spin-trapping reaction. In this, the trapping of alkyl radicals formed on thermolysis of the bis-heptenoyl peroxide (**5**) by MNP was studied. Both spin adducts, [hexenyl-MNP·] and [cyclopentylmethyl-MNP·], were formed (Scheme 5.3). Measurement of the relative proportions of these, coupled with a knowledge of the concentration of MNP and of the rate constant for cyclisation of the hexenyl radical (see above), permitted the rate constant for trapping of the primary hexenyl radical by MNP to be estimated (ca. 9×10^6 M^{-1} s^{-1} at 300 K).[1] When a reaction of known rate constant is used in this fashion to estimate the rate of some competing process, it has been referred to as a "clock" reaction. This terminology has been used particularly where the reference (clock) reaction is a unimolecular rearrangement, and clock reactions with a very wide range of time scales have been devised. An important application of the radical clock principle to an investigation of the mechanism of an enzymic oxidation process is described in Section 10.2.

Since the persistence of nitroxides underlies the success of spin trapping, we shall dwell for a moment on the structure and properties of these radicals.

[1]The experiment was less straightforward than described here, since both spin adducts are of the form $RCH_2N(Bu^t)O\cdot$, and therefore have almost identical (and exactly superimposed) ESR spectra, so that quantitative comparison of their concentrations is impossible. To circumvent this, ^{13}C was incorporated (~100%) at C-6 of the hexenyl radical, thus introducing an additional splitting into the spin-adduct spectrum from the cyclised radical and enabling it to be cleanly differentiated from the spectrum of the hexenyl adduct.

1 cal = 4.184 J

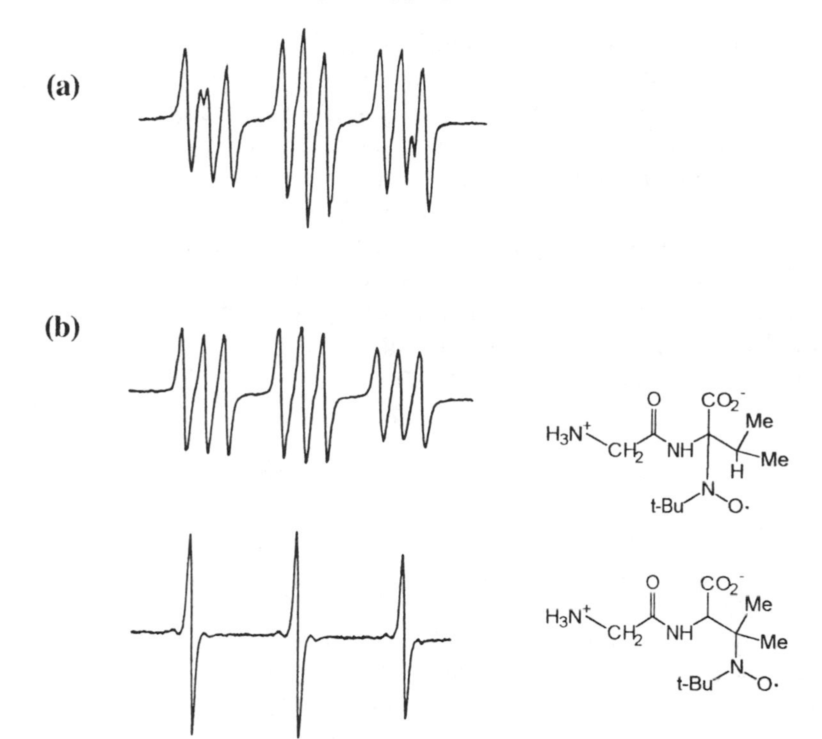

Fig. 5.12: The ESR spectra of spin adducts generated by MNP-trapping of radicals derived by hydrogen abstraction from the dipeptide glycylleucine, shown (**a**) before, and (**b**) after hplc separation. This experiment was carried out by radiolysis of crystalline dipeptide, followed by dissolution of the radiation-damaged crystals in an aqueous solution of MNP. (Reproduced with permission from K. Makino and P.B. Reisz, *Canad. J. Chem.*, **60**, 1480 (1982)).

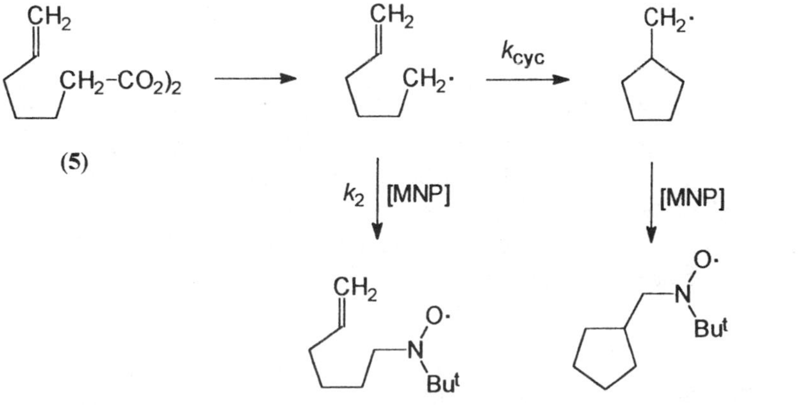

CH_2 CH_2–$CO_2)_2$ **(5)** → CH_2 $CH_2\cdot$ $\xrightarrow{k_{cyc}}$ $CH_2\cdot$

k_2 [MNP] [MNP]

CH_2 O· N Bu^t O· N Bu^t

Scheme 5.3

[hexenyl-MNP·] [cyclopentylmethyl-MNP·]

They have a superficial resemblance to two familiar inorganic paramagnetic species, NO and NO_2. The NO group of a nitroxide may be represented as a hybrid of the two resonance structures (**6a**) and (**6b**); the implied resonance stabilisation[1] is reflected in the weakness of the OH bond in hydroxylamines (see Table 4.3). An alternative, molecular orbital, picture of nitroxide structure places two electrons in an NO π-bonding orbital and one in the corresponding antibonding orbital, as well as two electrons in an NO σ-bond. The result is an approximate NO bond order of 1.5. An interesting consequence of this electronic structure is a particularly "soft" geometry at nitrogen – where it has been calculated that very little energy is required to deform the molecule from its preferred planar geometry into a fully tetrahedral (sp^3) structure.

$$>\ddot{N}-O\cdot \longleftrightarrow >\dot{N}^{+}-O^{-}$$

(**6a**) (**6b**)

Many nitroxides, e.g. porphyrexide (see p. 1), may be isolated and are indefinitely stable. Others decay more or less rapidly, depending upon the nature of the substituents on nitrogen. The methyl spin adduct of MNP, t-butyl methyl nitroxide, is in this latter category, and disproportionates to a hydroxylamine and nitrone [Equation (5)]. However, even when nitroxides are not stable – in the sense that they cannot be isolated – their rates of bimolecular decay are usually well below the diffusion limit, so that they too conform to our description "persistent" (Chapter 4). It will always be easier to establish detectable concentrations of persistent radicals than of species whose bimolecular self-reaction occurs at, or close to the diffusion-controlled rate.

$$2\ (Bu^t)(Me)N-O\cdot \longrightarrow Bu^t\text{-}N{=}O + (Bu^t)(CH_2{=})N^{+}-O^{-} \qquad (5)$$

Amongst the nitroxides which can be isolated are those which carry two tertiary alkyl groups on nitrogen, exemplified by $(t\text{-}Bu)_2NO\cdot$ and the tetramethyl-piperidone-*N*-oxyl (**7**). These nitroxides normally exhibit simple three-line ESR spectra, the only resolved hyperfine splitting arising from nitrogen. We shall see, however, in Chapter 10 that there may be subtle variations in these spectra which enable nitroxides to be used as "reporter groups" whose spectra give information on their

[1]These radicals are both persistent *and* stabilised (see Chapter 4, p. 47).

1 cal = 4.184 J

microscopic environment in a way that has been exploited in biological investigations.

(7)

5.5 ELECTRON NUCLEAR DOUBLE RESONANCE SPECTROSCOPY

Whilst several computational strategies have been developed for the analysis and interpretation of complex ESR spectra such as that of the triphenylmethyl radical, there is an experimental approach which can be particularly helpful, which involves a double resonance technique. In Electron Nuclear DOuble Resonance (ENDOR) spectroscopy the magnetic field and microwave frequency are adjusted to one of the ESR resonance lines, using high microwave power to partially saturate the signal. The radiofrequency corresponding to nuclear resonance at the applied magnetic field is then swept, and the intensity of the ESR signal is seen to respond when the nuclear resonance condition is met for each set of equivalent protons (or other magnetic nuclei). Thus for the six *meta* protons of triphenylmethyl, shown in Fig. **5.13**, the ENDOR spectrum exhibits two lines equally spaced on either side of the free proton resonance frequency. The spacing gives a measure (in frequency units) of the ESR proton hyperfine splitting. The same is true for the more strongly coupled *ortho* and *para* protons, except that only the high frequency lines are shown in the figure.

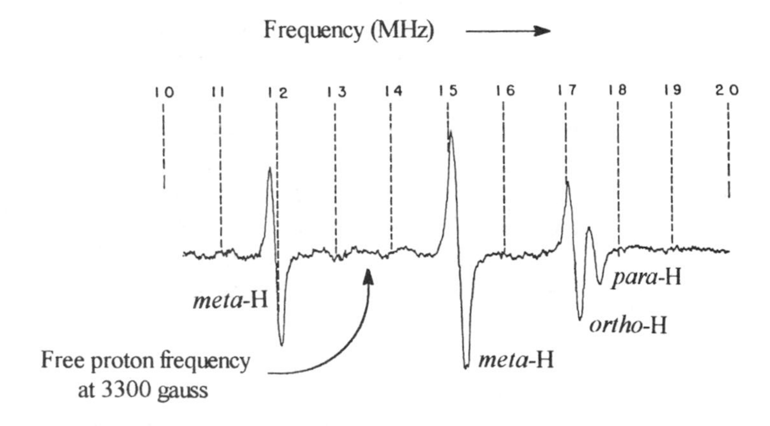

Fig. 5.13: The ENDOR spectrum of the triphenylmethyl radical. The simple pattern is to be contrasted with the ESR spectrum of Fig. **5.4**. (Reproduced with permission from J.S. Hyde, *J. Chem.Phys.*, **43**, 1806 (1965)).

Unfortunately, the technique is seldom helpful in those circumstances in which a complicated ESR spectrum is also very weak, as may occur when complex radicals are generated as reaction transients whose bimolecular decay is diffusion-controlled. For such species the radical concentration may be comparable to that of other relatively easily detected transients, such as methyl, but instead of just four lines in the spectrum, the same integrated signal intensity may be distributed between many tens or even hundreds of lines. An example would be the benzyl radical with a possible 54 distinct lines. The reason that ENDOR has been of little help in these circumstances is that it is generally significantly less sensitive than conventional ESR spectroscopy, frequently by as much as two orders of magnitude. Therefore ENDOR spectra of transients have been difficult to obtain.

5.6 CHEMICALLY INDUCED DYNAMIC NUCLEAR POLARISATION (CIDNP)

Experiments performed in the mid-1960s, in which reactions were carried out in sample tubes located in the probe region of an NMR spectrometer, occasionally produced what were, when first experienced, totally unexpected results. In these exceptional cases, some of the signals in spectra *which were recorded during the progress of a reaction* were found to be very much more intense than would be possible for 100% product yield. Even more surprisingly, other signals were sometimes in *emission* (i.e. peaks were inverted). In other examples, individual lines of a single multiplet would exhibit opposite effects. All of the reactions which showed this odd behaviour had one feature in common. They proceeded *via* radical intermediates. Two examples are given in Fig. **5.14**. Evidently the normal population of nuclear spins in the product molecules is being disturbed in these experiments – hence the designation "nuclear polarisation". Only after these reactions are effectively complete do the spin populations relax to their normal Boltzmann values, determined by the strength of the magnetic field; the spectra then exhibit normal intensities, appropriate to the concentrations of the different products.

A full explanation of these phenomena is beyond the scope of this text (and indeed it is of interest to recall that the earliest interpretation was in error). The following description is designed to give the reader a qualitative insight into how such effects might arise.

CIDNP phenomena have all been explained in terms of the behaviour of geminate radical pairs (see Chapter 3).[1] There is weak electron correlation between the unpaired electrons of the two radicals, so that we may speak of triplet or singlet radical pairs depending upon whether the two electron spins are parallel or

[1]The phenomena of CIDNP were first properly understood, and were put on a quantitative footing by Gerhard Closs in the United States and, independently, by Kaptein and Oosterhoff in Holland in 1969.

1 cal = 4.184 J

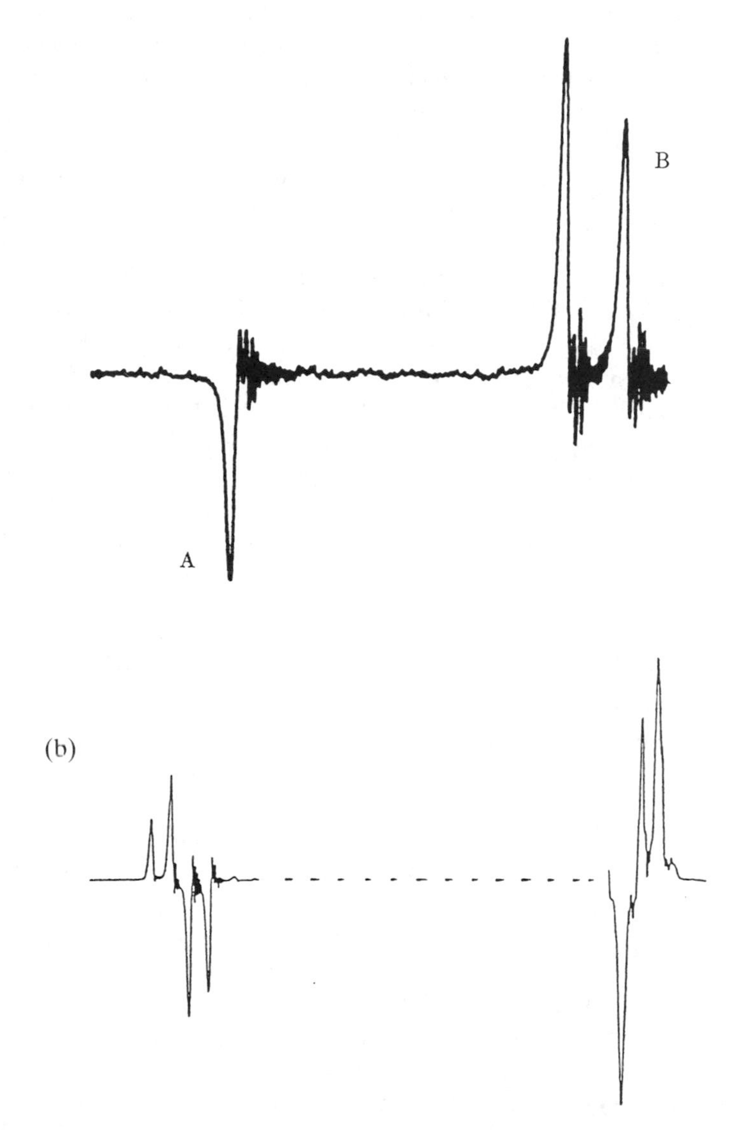

Fig. 5.14: NMR spectra recorded during the progress of two peroxide decompositions. (**a**) Partial spectrum of the reaction mixture during the decomposition of acetyl trichloroacetyl peroxide in the presence of iodine: the peak marked A, in emission, is due to the cage recombination product CH_3CCl_3; that marked B, showing enhanced absorption, is due to the escape product, methyl iodide. The third signal in the spectrum was not unambiguously identified. (Reproduced with permission from H.R. Ward, *Accounts Chem. Res.*, **5**, 18 (1972)). (**b**) Partial spectrum of ethyl phenyl selenide formed by S_H2 displacement on selenium during the decomposition of dipropionyl peroxide in the presence of diphenyl diselenide. The quartet and triplet features due to the ethyl protons show clear multiplet polarisation. (Reproduced with permission from the Ph.D.Thesis of P.B. Bishop, University of London, 1985).

antiparallel respectively. Thermolysis of a peroxide or of an azo-compound will initially generate a singlet pair, rupture of a bond being faster than spin inversion; some photochemical reactions will generate triplet pairs,[1] and, in the more general case of diffusive encounter, radical pairs will be formed as a (quantum) statistical mixture (triplet:singlet = 3:1). For the radical pair to collapse to diamagnetic material by coupling or disproportionation, the partners must be in the singlet state. Supposing the pair starts predominantly in the spin-paired (singlet) form, spin inversion will facilitate diffusive separation. This will occur at a rate dependent upon the magnetic properties of the individual radicals within the pair. In particular, if the radicals differ slightly in *g*-value the unpaired electrons will precess at slightly different frequencies [see Equation (1)], thus permitting singlet–triplet crossing to occur. The rate of spin inversion will also be modulated by the hyperfine field due to any magnetic nuclei present in one of the two radicals: nuclei with spin state of one sign in one of the radicals might enhance the rate of intersystem crossing due to *g*-value differences; those of opposite sign would by contrast attenuate it. Diamagnetic products formed by spin inversion followed by diffusive separation and subsequent reaction of the free radicals, e.g. with solvent, will be enriched in nuclei of the spin state which facilitated electron spin inversion, whilst products formed by collapse of the radical pair will be enriched in nuclei of the spin state which hindered inversion. This is nicely illustrated by results obtained for the [$CH_3\cdot$ $CCl_3\cdot$] pair generated by decomposition of the unsymmetrical peroxide CH_3CO-O-O-$COCCl_3$ in benzene containing dissolved iodine. The protons of 1,1,1-trichloroethane, the cage recombination product, initially appear as an intense emission signal, whilst the protons of iodomethane formed following escape from the solvent cage are in enhanced absorption (Fig. **5.14a**).

As little as 1% spin polarisation can lead to signal enhancements of 1000-fold, so that the phenomenon may be very sensitive to radical participation. It should be emphasised, therefore, that whilst observation of CIDNP effects is unambiguously indicative of radical intermediates, these may contribute to only a tiny fraction of the overall chemical change in the system under examination.

When the initial radical pair is formed in the triplet state, opposite arguments determine the sign of polarisation in recombination and escape products. Here, spin inversion must precede geminate recombination.

Even when the *g*-values of both radical partners in the geminate pair are identical, nuclear spin effects may reveal themselves – in so-called "multiplet effects". This is illustrated in Fig. **5.14b**, which shows polarisation in the ethyl protons of ethyl phenyl selenide formed during decomposition of dipropionyl peroxide (EtCO.OO.COEt) in the presence of PhSeSePh.

[1]For example when a photo-excited triplet species, ^{3}**S*** (e.g. benzophenone), abstracts hydrogen from a ground-state molecule (^{3}**S*** + **M-H** → ^{3}**[S-H· M·]**); see also Chapter 9.

1 cal = 4.184 J

One of the consequences of the effect of nuclear spin on radical-pair behaviour is the possibility of influencing the partitioning of pairs between cage recombination and escape. One ingenious application of this principle has been to the nuclear-spin based fractionation of compounds according to their isotopic content. For example ^{12}C and ^{13}C at the reaction centre will be magnetically differentiated; the macroscopic effect of this can be magnified very significantly by enhancing the pair life-time by means of micellar encapsulation (see the bibliography).

Nuclear spin relaxation times are considerably longer than the corresponding electron spin values, but nevertheless in some ESR experiments it has been possible to detect abnormal line intensities that can be attributed to residual electron spin polarisation. The phenomenon has been designated CIDEP (pronounced "sci-dep") – chemically induced dynamic electron polarisation; it has been extensively studied using time-resolved ESR techniques.

6

Radicals and radical reactions in organic synthesis

6.1 INTRODUCTION

In Chapter 1 we noted that elementary organic chemistry courses were likely to have introduced the student to radical coupling reactions – perhaps through the synthetic utility of the Kolbe electrochemical oxidation (Scheme 1.3) – and to the importance of chain reactions – probably by a simple exposition of alkane halogenation. More interesting examples of coupling may later have been encountered in the guise of one-electron reductions of ketones [e.g. the pinacol and McMurry reactions; Equations (1) and (2)] and of esters (e.g. the acyloin reaction; Scheme 6.1). Significant chain reactions that might have been added to halogenation could have

Pinacol,

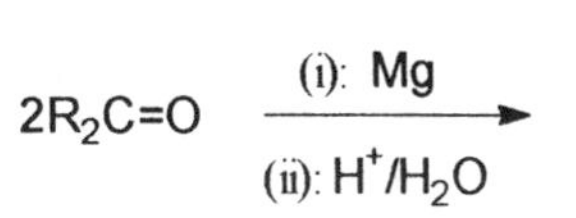

$$2R_2C{=}O \xrightarrow[\text{(ii): } H^+/H_2O]{\text{(i): Mg}} R_2C(OH){-}C(OH)R_2 \qquad (1)$$

and McMurry reactions,

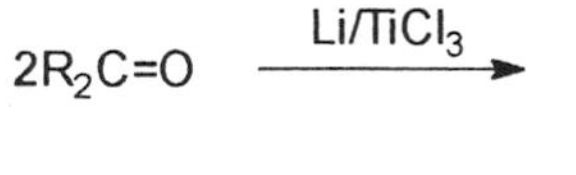

$$2R_2C{=}O \xrightarrow{Li/TiCl_3} R_2C{=}CR_2 \qquad (2)$$

both *via* dimerisation of $R_2C{=}O^{\bullet-}$

included the radical pathway for addition polymerisation, and perhaps the very important tributyltin hydride reduction of (particularly alkyl) halides (Scheme 6.2). But Chapter 1 should also have tantalised any reader having even a passing interest in organic synthesis, by providing a glimpse of the far more complex

1 cal = 4.184 J

transformations that free-radical chemists have wrought in their molecules, particularly during the past 10–15 years. Because of space limitations, we shall for the moment pay little further attention to the many important one-electron redox reactions which, like the examples of Equations (1) and (2) and Scheme 6.1, are usually classified with oxidations and reductions. Exceptions will exemplify the use of these changes to initiate more typically radical processes. More information will, however, be found in Chapter 8.

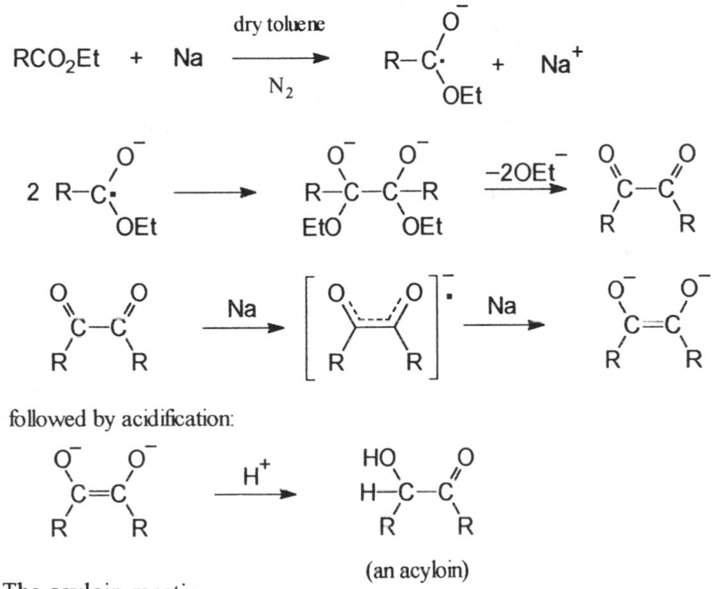

$$RCO_2Et + Na \xrightarrow[N_2]{\text{dry toluene}} R{-}\dot{C}(O^-)(OEt) + Na^+$$

$$2\ R{-}\dot{C}(O^-)(OEt) \longrightarrow R{-}C(O^-)(EtO){-}C(O^-)(OEt){-}R \xrightarrow{-2OEt^-} (O{=})C(R){-}C(R)({=}O)$$

$$(O{=})C(R){-}C(R)({=}O) \xrightarrow{Na} \left[O{\cdots}C(R){\cdots}C(R){\cdots}O \right]^{\cdot -} \xrightarrow{Na} (^-O)C(R){=}C(R)(O^-)$$

followed by acidification:

$$(^-O)C(R){=}C(R)(O^-) \xrightarrow{H^+} H{-}C(HO)(R){-}C(R)({=}O)$$

(an acyloin)

Scheme 6.1: The acyloin reaction.

Overall reaction:

$$Bu_3SnH + RX \longrightarrow RH + Bu_3SnX$$

Chain propagating steps:

$$Bu_3SnH + R\cdot \longrightarrow RH + Bu_3Sn\cdot$$

$$Bu_3Sn\cdot + RX \longrightarrow Bu_3SnX + R\cdot$$

Scheme 6.2

6.2 CHAIN REACTIONS IN SYNTHESIS

Functional group interconversions: Whilst chlorination of alkanes has been seen to be unselective, it is important to recall the role of molecular symmetry. Thus, since all of the hydrogens of cyclohexane are equivalent, there is only one monochlorocyclohexane, and chlorination of this hydrocarbon is synthetically viable. This symmetry factor has been put to good use in the Toray process, developed in Japan, for generating caprolactam (a precursor to Nylon-6) *via* the photonitrosation of cyclohexane using nitrosyl chloride. In that reaction, hydrogen abstraction is by chlorine atoms. Interestingly, this is a non-chain process (Scheme 6.3), and therefore requires considerable power input, yet it is so efficient that it has competed commercially with other routes from cyclohexane to caprolactam.

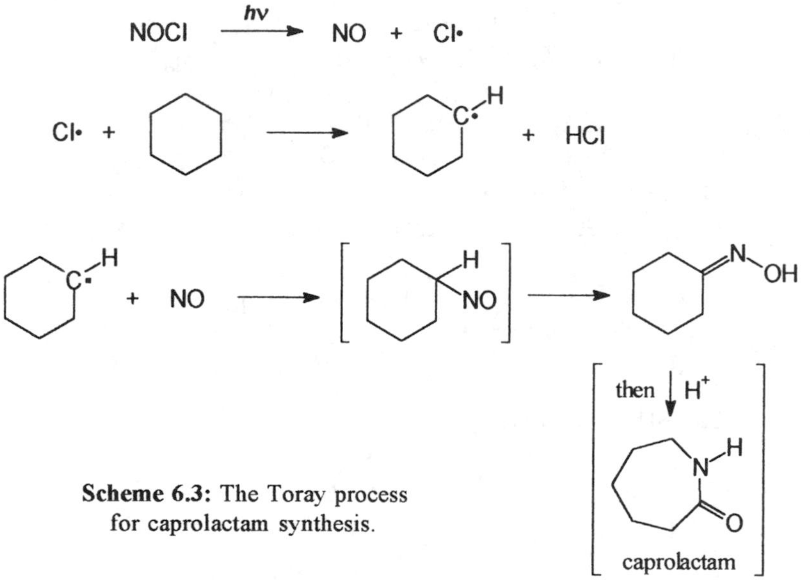

NOCl —hν→ NO + Cl•

Cl• + cyclohexane → cyclohexyl radical + HCl

cyclohexyl radical + NO → [nitrosocyclohexane] → cyclohexanone oxime

then H⁺ → caprolactam

Scheme 6.3: The Toray process for caprolactam synthesis.

A laboratory reagent which functionalises hydrocarbons *selectively* by hydrogen abstraction is *N*-bromosuccinimide (NBS). With this, only relatively weakly held benzylic or allylic hydrogens are replaced (by bromine); for allylic bromination it must be recognised that, because of the intermediacy of allylic radicals, the reaction may be accompanied by double-bond migration (Scheme 6.4). Other aspects of the mechanism of allylic bromination by NBS were discussed in Chapter 4.

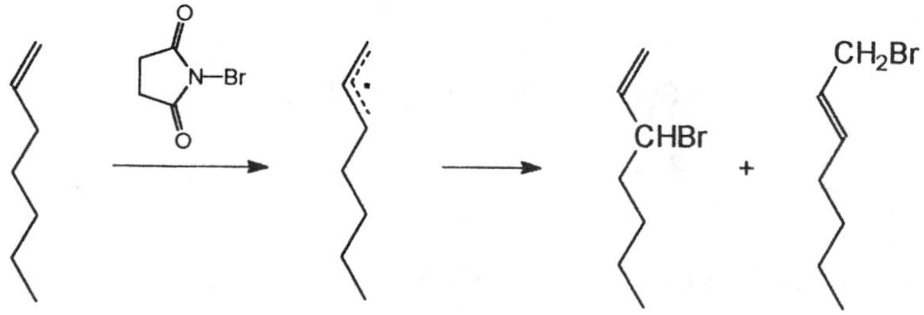

1-octene —NBS→ allylic radical → CHBr product + CH_2Br product

Scheme 6.4

1 cal = 4.184 J

Benzylic brominations may also be effected using bromotrichloromethane (in which chain propagation includes hydrogen abstraction by trichloromethyl radicals), and benzylic (and other) chlorinations by t-butyl hypochlorite (Bu^tOCl, in which case the abstraction is by t-butoxyl radicals).

Another important aspect of hydrocarbon functionalisation is that in which selectivity is achieved by proximity effects in intramolecular reactions. Protonated aminyl radicals (unlike their non-protonated counterparts) are sufficiently reactive to abstract hydrogen from saturated hydrocarbons. This fact has been utilised in photoinitiated reactions of protonated *N*-chloramines in which an *N*-alkyl substituent has a δ-hydrogen, e.g. (**1**). The resulting δ-halogenated amines are not normally isolated. Instead, cyclisation to pyrrolidines is promoted by addition of base. This is the Hoffman–Löffler–Freytag reaction, which has found particular application in alkaloid synthesis.

Initiation:

H N^+HMeCl (1) $\xrightarrow{h\nu}$ H $\dot{N}^+HMe$

Propagation:

H $\dot{N}^+HMe$ ⟶ • N^+H_2Me

• N^+H_2Me + (1) ⟶ Cl N^+H_2Me + • N^+H_2Me

Base

NMe

In steroid and triterpene chemistry, functionalisation of unactivated axial methyl groups by hydrogen transfer to an axial alkoxyl radical at C-11 is exemplified by Barton's synthesis of aldosterone (**3**). This was accomplished by photolysis of the nitrite ester (**2**). At the time (1960), this constituted a major breakthrough in synthetic strategy. It was over a decade before the more elaborate achievements of remote functionalisation in steroid reactions, typified by the transformation shown in Scheme 6.5, were developed by Breslow. Scheme 6.5 not only illustrates the concept of remote functionalisation – in which the aryl ester acts as a template which holds the reaction centre at an appropriate distance from the anchorage point

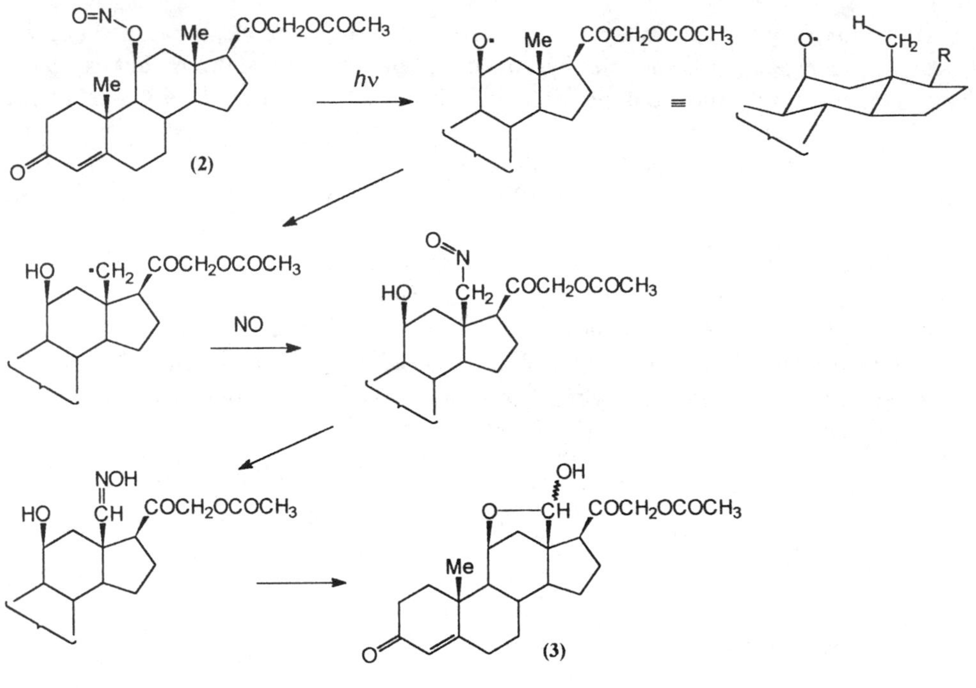

O=N O Me COCH$_2$OCOCH$_3$ Me (2) O $h\nu$ O• Me COCH$_2$OCOCH$_3$ ≡ O• H CH$_2$ R

HO •CH$_2$ COCH$_2$OCOCH$_3$ NO O N HO CH$_2$ COCH$_2$OCOCH$_3$

NOH HO CH COCH$_2$OCOCH$_3$ OH O CH COCH$_2$OCOCH$_3$ Me O (3)

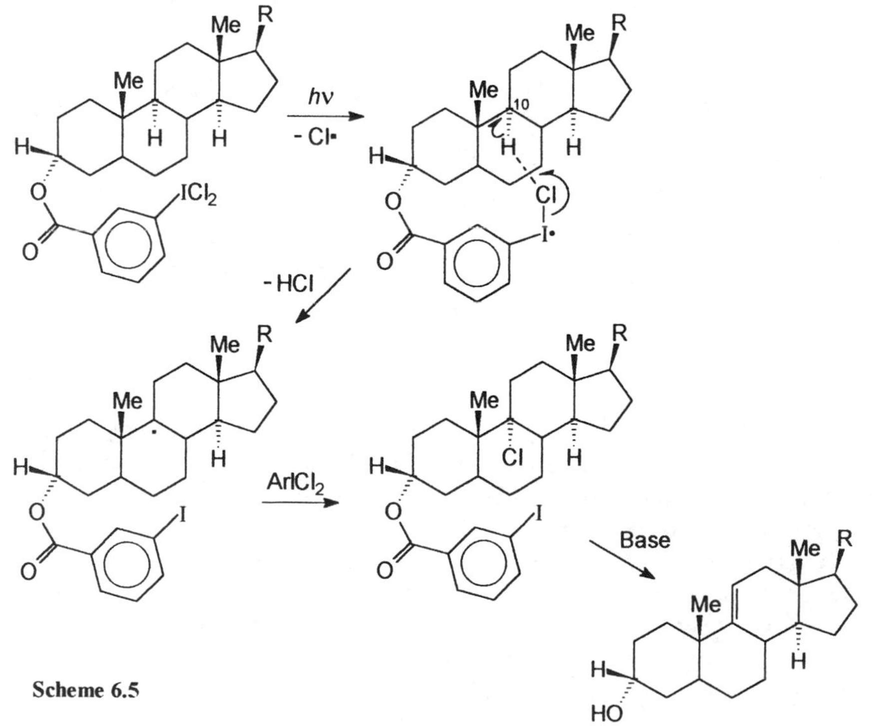

Me R Me H H H O ICl$_2$ O $h\nu$ - Cl• Me R Me 10 H H H O Cl I• O

- HCl

Me R Me H H O I O ArICl$_2$ Me R Me Cl H H O I O Base Me R Me H H HO

Scheme 6.5

on C-3 to permit selective abstraction of the tertiary hydrogen at C-10[1] – it also relates to one of the few well-documented examples of solvent effects in radical chemistry. Thus, in Chapter 4, we discussed the very low selectivity exhibited by chlorine atoms in hydrogen abstraction (per hydrogen tertiary/primary reactivity ratio ≈ 5). If alkane chlorinations are carried out in the presence of an aromatic diluent, e.g. benzene, the selectivity increases markedly (tertiary/primary ≈ 20 in the presence of 4 M benzene). It has become clear that the explanation for this is that the chlorine's reactivity is attenuated by association with the π-system of the benzene ring, since the magnitude of the effect is related to the ionisation potential of the aromatic solvent. However, in the case of iodobenzene the effect is especially pronounced, and it has been concluded that it is then the PhI(Cl)· radical which is the hydrogen abstractor, as in the intramolecular example involving photolysis of Breslow's dichloroiodo-compound.

One of the most significant recent developments in simple functional group interconversion by free-radical methodology is that developed by the team led by Barton. Starting with aliphatic carboxylic acids, replacement of the carboxyl group by a wide variety of substituents (e.g. H, Hal, SPh, SePh, PO_3H, OOH, OH) has been accomplished, usually with good to excellent yields. The method, which is both more versatile than, and generally superior to, the reaction of silver carboxylates with halogens (the Hunsdiecker reaction), depends on initial formation of an *O*-acyl derivative of *N*-hydroxy-2-pyridinethione (**4**). Possible interconversions are illustrated in Scheme 6.6; the chain-propagating steps for two

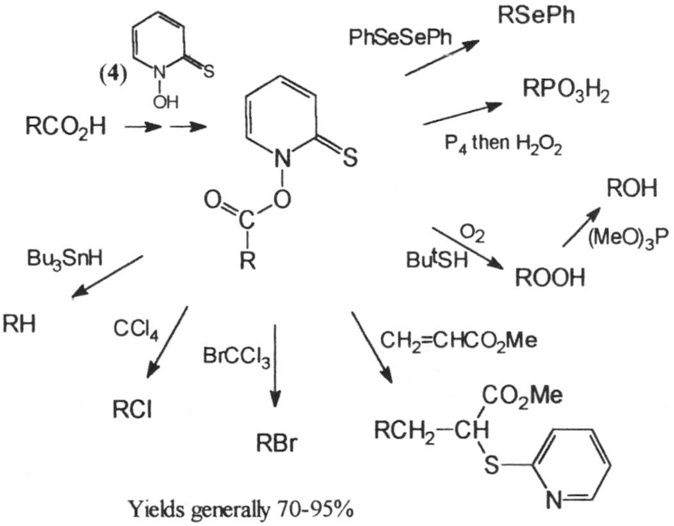

Scheme 6.6

[1]Creation of a C-10–C-11 double bond affords a means of introducing oxygen functionality to C-11, characteristic of cortisone and other steroidal hormones of the adrenal cortex.

representative transformations are given below. These transformations may be initiated thermally, e.g. by AIBN (Chapter 3), or photolytically, although in some instances reaction occurs without the necessity for active initiation. Where a thermal initiator is necessary, a useful general point to remember is that better yields are normally obtained if the initiator is added in portions during the progress of the reaction, and not all at once at the outset.

(note *bridgehead* functional group change)

Another development is a procedure for the deoxygenation of alcohols, which is of especial significance in carbohydrate chemistry. Shown in Scheme 6.7(a), this depends on the ready addition of a trialkyltin radical to the sulphur of an *O*-alkyl thioester or xanthate, and the fragmentation of the adduct radical. The reaction occurs at ca. 100°C when R is secondary, but significantly higher temperatures are required when R is primary. Because of the problems associated with derivatisation of tertiary alcohols, these are first converted into the corresponding half ester, half acid chloride of oxalic acid and thence into a derivative of **(4)**, as indicated in Scheme 6.7(b).

1 cal = 4.184 J

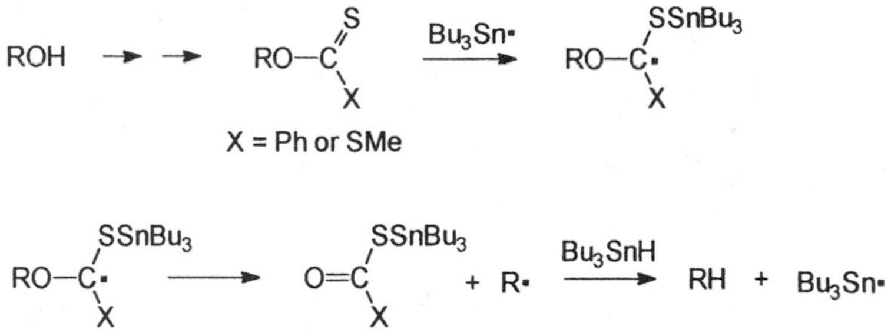

$$ROH \rightarrow \rightarrow RO{-}C({=}S){-}X \xrightarrow{Bu_3Sn\cdot} RO{-}\dot{C}(SSnBu_3){-}X$$

X = Ph or SMe

$$RO{-}\dot{C}(SSnBu_3){-}X \longrightarrow O{=}C(SSnBu_3){-}X + R\cdot \xrightarrow{Bu_3SnH} RH + Bu_3Sn\cdot$$

Scheme 6.7(a)

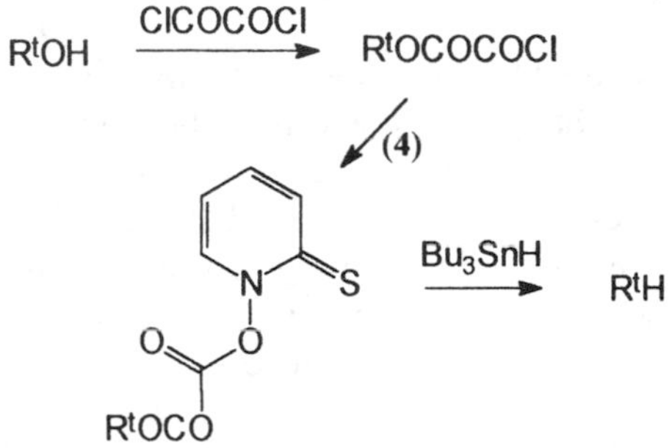

$$R^tOH \xrightarrow{ClCOCOCl} R^tOCOCOCl$$

(4)

$$R^tOCOC(O)O{-}N\text{(pyridine-2-thione)} \xrightarrow{Bu_3SnH} R^tH$$

Scheme 6.7(b)

Carbon–carbon bond formation: Anti-Markownikov functionalisation of alkenes by HBr addition (Chapter 1) has in general been superseded as a synthetically useful reaction by hydroboration, but numerous other radical additions to alkenes are preparatively useful, in particular those which generate new carbon–carbon bonds. Such reactions are many and varied. Whilst some may be complicated by competing telomer[1] production, there are many for which the rate constants of individual steps allow 1:1-adducts to be isolated in excellent yields. Examples, all of which give yields in excess of 70%, are presented in Scheme 6.8. The best yields

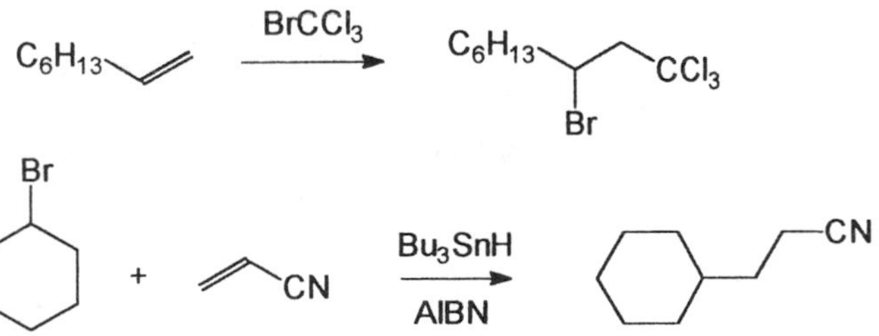

$$C_6H_{13}CH{=}CH_2 \xrightarrow{BrCCl_3} C_6H_{13}CHBrCH_2CCl_3$$

$$\text{cyclohexyl-Br} + CH_2{=}CHCN \xrightarrow[AIBN]{Bu_3SnH} \text{cyclohexyl-}CH_2CH_2CN$$

Scheme 6.8 *(continued overleaf)*

[1]A telomer is a short-chain polymer and may contain as few as two alkene units.

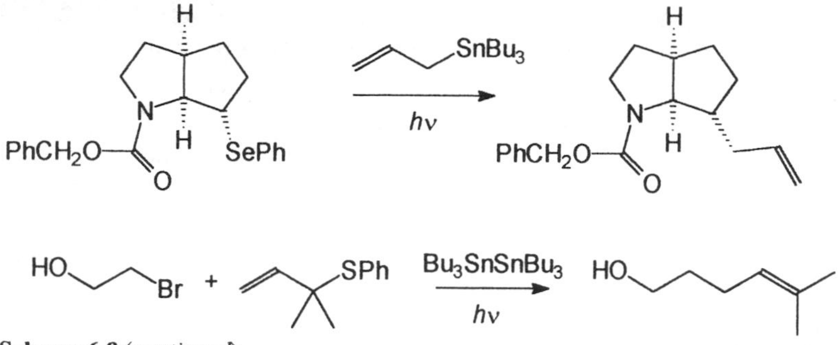

Scheme 6.8 (*continued*)

from additions of simple alkyl radicals (which are weakly nucleophilic; see discussion of polar effects in Chapter 4) are often obtained with electrophilic alkenes, such as acrylonitrile or methyl acrylate, and with alkyl bromides or iodides.[1] The final product-forming step in a radical addition cycle is also influenced by polar effects. For example, in the cyanoethylation reaction the intermediate cyanoethyl radical is rendered relatively electrophilic by the cyano substituent. Its tendency to add to a further molecule of acrylonitrile is thereby diminished when compared with the initial alkyl radical, whereas its attack on the hydrogen atom bonded to the electropositive tin atom is facilitated. The propagating steps in this reaction are elaborated in Schemes 6.9(a) and 6.9(b). The latter format, reminiscent of schemes representing enzymic catalysis, has been routinely adopted

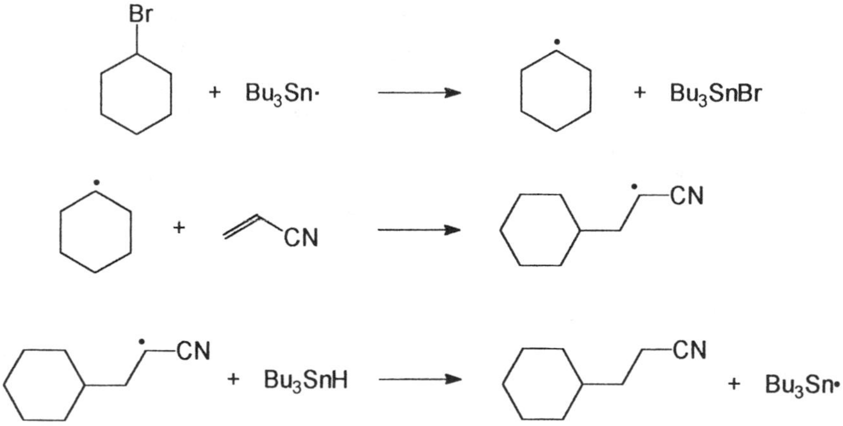

Scheme 6.9(a)

[1]Typically, reactions of alkyl radicals with methyl acrylate may be some three orders of magnitude faster than with a simple terminal alkene.

1 cal = 4.184 J

by some authors. In this reaction, the product may be viewed as the adduct of cyclohexane and acrylonitrile, and it might be asked whether the same compound can be produced by direct radical reaction between these two compounds. Although modest yields of 1:1 adducts *have* been achieved in some C-H additions to alkenes, a key step involves a hydrogen-atom transfer between two carbon atoms, for which

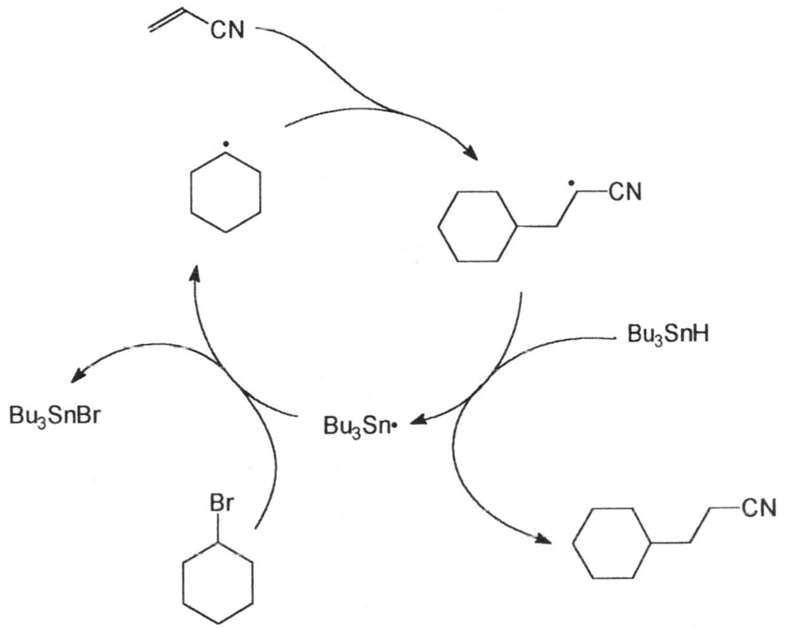

CN

CN

Bu_3SnH

Bu_3SnBr

$Bu_3Sn\cdot$

Br

CN

Scheme 6.9(b)

the activation barrier is relatively high. Competing telomer production is therefore a serious complication. Furthermore, in the example of acrylonitrile and cyclohexane, inspection of data in Table 4.3 would suggest that one essential step [Equation (3)] would be appreciably endothermic. In contrast, the tin hydride–alkyl bromide reaction is rapid and creates a specific alkyl radical centred at the site originally bearing the halogen atom.

CN + → CN + (3)

The reaction of tributyltin radical with alkyl chlorides is relatively slow, so that competing addition of the tin radical to the alkene may be observed. On the other hand, alkyl phenyl selenides are satisfactory alternatives to alkyl halides: S_H2 substitution at selenium by the tributyltin radical displaces the alkyl group rather than the more tightly bound phenyl.

The last two examples in Scheme 6.8 involve addition-elimination, in which a carbon-centred radical adds to an allylstannane or an allylic thioether. These C-allylation reactions are generally very efficient, although they are seldom successful when the terminal carbon of the double bond is substituted. Chain propagation in these two reactions is interesting. In the first, alkyl radical addition is followed by loss of the $Bu_3Sn\cdot$ radical which then reacts with the halogen compound. In the second, a thiyl radical is formed which reacts with the hexabutylditin to release a stannyl radical in a S_H2 process. Both the last example, and the related reaction shown on Page 5, illustrate again the fact that it is seldom necessary to protect alcoholic OH groups during radical processes. More generally, the pattern of reactivity of various functional groups towards radical intermediates is quite distinct from what is found with powerful nucleophiles or electrophiles. Importantly, therefore, many groups generally regarded as sensitive to transformation under ionic reaction conditions are unaffected in a radical reaction.

In an alternative procedure for effecting alkyl-radical addition to alkenes, the alkyl halide–tributyltin hydride combination has been replaced by an alkyl mercuric halide and sodium borohydride. This effects *in situ* generation of an alkyl mercuric hydride, RHgH, which undergoes radical chain addition of RH and liberation of mercury (e.g. Scheme 6.10). Although the yields are usually somewhat inferior to those obtained by the tin hydride method, the ready accessibility of organomercury compounds, in particular by mercuration of alkenes (either directly or *via* prior hydroboration – Scheme 6.11), frequently makes this alternative an attractive one.

C_6H_{11}–HgCl $\xrightarrow{NaBH_4}$ [C_6H_{11}–HgH]

followed by:

C_6H_{11}–HgH + CH_2=CHCN ⟶ C_6H_{11}–CH_2CH_2CN + Hg

via the propagating sequence:

$C_6H_{11}\cdot$ + CH_2=CHCN ⟶ C_6H_{11}–$CH_2\dot{C}HCN$

C_6H_{11}–$CH_2\dot{C}HCN$ + C_6H_{11}–HgH

⟶ C_6H_{11}–CH_2CH_2CN + $C_6H_{11}\cdot$ + Hg

Scheme 6.10

1 cal = 4.184 J

$$RCH{=}CH_2 \xrightarrow{Hg(O_2CCH_3)_2} CH_3CO_2CH(R)CH_2HgO_2CCH_3 \xrightarrow[CH_2{=}CHCN]{NaBH_4} CH_3CO_2CH(R)CH_2CH_2CH_2CN$$

$$RCH{=}CH_2 \xrightarrow{BH_3 \text{ followed by } Hg(O_2CCH_3)_2} RCH_2CH_2HgO_2CCH_3 \xrightarrow[CH_2{=}CHCN]{NaBH_4} RCH_2CH_2CH_2CH_2CN$$

Scheme 6.11

Yet another reaction type which involves radical addition exploits the application of high pressure to laboratory synthesis. By this means, it has been possible to achieve synthetically useful yields in radical additions to carbon monoxide (e.g. Scheme 6.12), an interesting reversal of the usual fragmentation of acyl radicals (Chapter 2). The reversibility of acyl decarbonylation has also been demonstrated in the oxidising environment of a Fenton system (Chapter 3); hydrogen abstraction from a (water-soluble) substrate R-H by HO· gives R· which adds reversibly to CO. The acyl radical so formed is rapidly intercepted by one-electron oxidation, and the resulting acylium ion reacts with water to give carboxylic acid, RCO_2H.

$$C_6H_{11}{-}I + Bu_3Sn\cdot \longrightarrow C_6H_{11}\cdot + Bu_3SnI$$

$$C_6H_{11}\cdot + CO \rightleftharpoons C_6H_{11}{-}\dot{C}{=}O$$

$$C_6H_{11}{-}\dot{C}{=}O + Bu_3SnH \longrightarrow C_6H_{11}{-}CHO + Bu_3Sn\cdot$$

Scheme 6.12: (CO pressure ca. 50 atmospheres)

Some of the most significant applications of radical reactions in synthesis depend on *intramolecular* additions. In particular, the hexenyl → cyclopentylmethyl rearrangement has provided a versatile approach to five-membered rings. This behaviour is in marked contrast to that of 5-hexenyl cations: solvolysis of hexenyl derivatives normally leads to cyclohexanes.

Three examples of hexenyl radical cyclisation are shown which illustrate the construction of triquinane skeletons. Numerous terpenoid compounds have been encountered which incorporate these tricyclic cyclopentanoid structures, and some, such as coriolin, have pronounced cytotoxic properties. Each of the examples given here, using precursors **(5)**–**(7)**, incorporates the concept of "tandem cyclisation" in which two rings are constructed in the same reaction, yet each of these examples proceeds in at least 50% yield with good stereocontrol. Each also exhibits other features of especial interest. For example the radical derived by transfer of the iodine atom from **(5)** to the tributyltin radical cyclises on to the cyclopentadiene to give a resonance stabilised allylic radical, as shown (Scheme 6.13). Nevertheless, addition of this stabilised intermediate to the exocyclic double bond is still sufficiently rapid that it can compete successfully with reduction by the tin hydride. At first sight, this result is all the more remarkable when it is recognised that the final product incorporates three contiguous quaternary carbons. However, quaternary centres are rather easily formed by addition of tertiary radicals to (especially electron-deficient) alkenes. The addition is strongly exothermic, and the C–C separation at the transition state is quite large (see Chapter 4) so that steric interference is poorly developed.

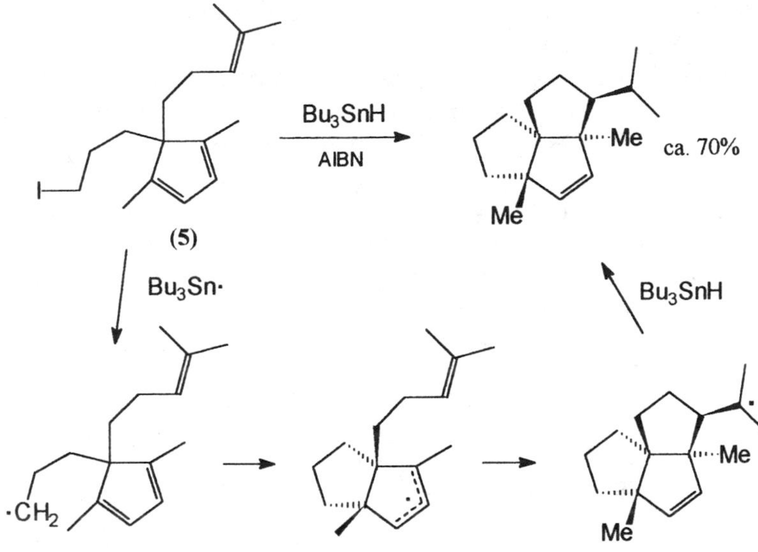

Scheme 6.13

In the second example, the initial radical is of the vinyl type. Although the bromine atom in **(6)** is removed from the position *trans* to the linkage to the cyclopentene, vinyl radicals undergo very rapid[1] configurational inversion, and the first cyclisation step is therefore unremarkable. This time, a secondary alkyl radical results, which cyclises without difficulty on to the exocyclic double bond.

The third example is interesting for the different approach to an initial odd-electron species, which here involves one-electron reduction of the aldehyde **(7)**, giving a ketyl radical ($RR'CO^{\cdot-}$) similar to those involved in pinacol and related reactions mentioned at the beginning of this chapter. In this case, the second cyclisation is of a hexynyl moiety. This has the considerable advantage, from the

[1]See Appendix.

1 cal = 4.184 J

standpoint of synthetic strategy, that the product incorporates an exocyclic double bond which is then available for further chemical manipulation.

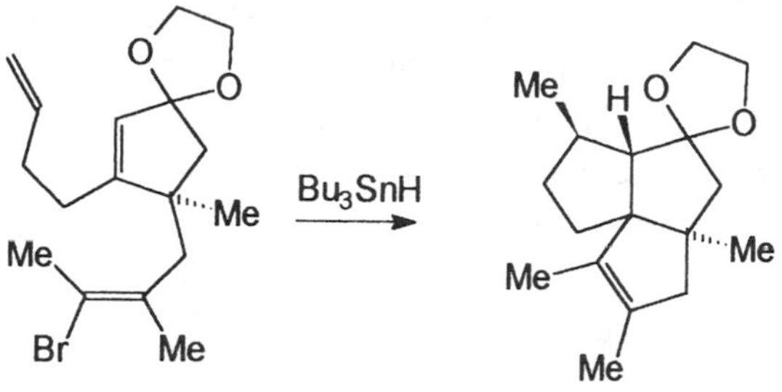

Me Br Me Me Me Bu_3SnH Me H Me Me Me Me

(6)

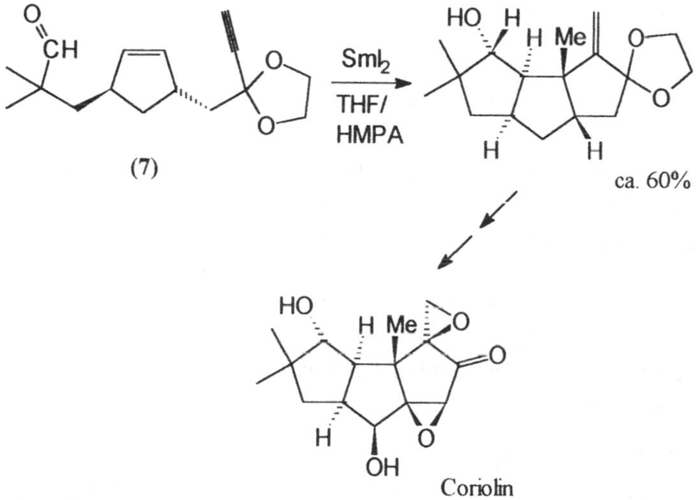

O CH SmI$_2$ THF/ HMPA HO H H Me H H

(7)

ca. 60%

HO H Me O O H OH O

Coriolin

In each of these three examples the stereochemistry of substituent attachment to the pre-formed five-membered ring determines the (*cis*) stereochemistry of ring-fusion. This overcomes the stereochemical complexity attendant on tandem cyclisations of radicals related to **(8)**. In that case two monocyclic intermediates can be formed, only one of which is able to cyclise further.

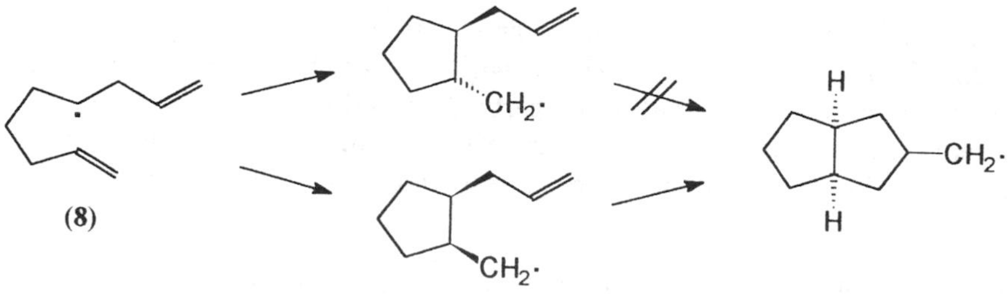

$CH_2\cdot$ H H $CH_2\cdot$ $CH_2\cdot$

(8)

With properly constructed precursors, it is evident that excellent yields have been achieved in some of these reactions, and inevitably this invites the question of whether or not we are limited to 5-membered rings. Many examples demonstrate that this need not be the case. In a synthesis of an analogue (**9**) of lysergic acid, the initial 5-*exo* cyclisation of an aryl radical[1] is followed by two 6-*endo* cyclisations, and the sequence ends with elimination of a PhS· radical. It is not entirely clear whether conformational strain tips the kinetic balance in favour of the 6-membered rings, or, as in certain other cases, ring-closure in this system is reversible, and it is the thermodynamically favoured cyclisation which is observed.

Bu_3SnH

(**9**; 20%)

A final example illustrates the extension to macrocyclic systems, with an efficient route to zearalenone (**10**). This result may seem all the more astonishing when it is recognised that the cyclising radical is both allylic and further stabilised by the aryl substituent, although, as with the allylic intermediate derived from (**5**), these factors will not only slow the cyclisation but also the competing intermolecular reduction. However, in the macrocyclisation reaction success depends on the presence of an electron-deficient alkene as the radical acceptor; without the carbonyl group, cyclisation fails.

Bu_3SnH

(**10**)

Many of the examples in this chapter utilise organotin hydride chemistry, introduced in Scheme 6.2. There will always be a competition between intramolecular addition of the initial carbon-centred radical and its intermolecular interception by R_3SnH before cyclisation can occur. To avoid substantial direct reduction without cyclisation, it is often essential to keep tin hydride concentrations

[1] This example also illustrates the possibility of generating aryl radicals by the tin hydride method, despite the relatively high strength of aryl–halogen bonds.

1 cal = 4.184 J

very low. In some instances this can be achieved satisfactorily only by very slow addition of tin hydride and initiator,[1] perhaps over a period of hours, using syringe-pump techniques.

A further problem with tin hydrides is the difficulty of removing residual tin compounds from reaction mixtures. This may constitute a serious manipulative problem,[2] but also, because of the toxicity of tin compounds, the methodology is effectively excluded from many practical applications – e.g. the preparation of pharmaceutical intermediates.[3] Two interesting solutions to this problem have been suggested. In the first, tributyltin hydride is replaced by tris(trimethylsilyl)silane. The Si–H bond in this compound is much weaker than that in, for example, trimethylsilane (Table 4.3), and, unlike the latter, ready reaction occurs with many alkyl halides in a radical-chain dehalogenation process.

Arguably more ingenious still is replacement of the stannane by a mixture of a simple trialkylsilane and a catalytic quantity of an aliphatic thiol. Use of trialkylsilane alone fails because of the relatively slow hydrogen transfer to alkyl radicals, but the more electrophilic hydrogen of the thiol is easily removed by the nucleophilic alkyl radical. In its turn, the thiyl radical is relatively electrophilic, and reacts rapidly with the trialkylsilane. This behaviour, summarised in Scheme 6.14, has been designated "polarity reversal catalysis"; the importance of the polar effect here is made more obvious when it is remembered that the S–H bond is even stronger than the Si–H bond (Table 4.3).

$$R\cdot + R'SH \longrightarrow R'S\cdot + RH$$

$$R'S\cdot + R''_3SiH \longrightarrow R'SH + R''_3Si\cdot$$

$$R''_3Si\cdot + RX \longrightarrow R''_3SiX + R\cdot$$

Scheme 6.14

[1]The preferred thermal initiator, particularly in tin hydride reactions, is normally AIBN. Slightly higher initiation temperatures can be achieved with the related azo-compound (**i**). Peroxide initiators may be rapidly reduced by organotin hydrides.

NC, N=N, CN

(i)

[2]Amongst the protocols adopted to remove tin residues, a widely recommended procedure involves dilution with undried ether and then titration with iodine to oxidise hexaalkylditin to trialkyltin iodide, followed by addition of diazabicycloundecene, which reacts with the tin halides. The mixture is then filtered through silica gel (*J. Org. Chem.*, 1989, p. 3140).

[3]This stricture is, of course, also pertinent to the mercuric hydride procedures.

6.3 NON-CHAIN PROCESSES

Most of the reactions which have been discussed above have chain mechanisms, and although there is undoubtedly a conceptual elegance and simplicity associated with the use of chain reactions for accomplishing complex synthetic transformations, the existence of important non-chain processes in which stoichiometric quantities of radical source are used must not be overlooked. One of the oldest involves the use of the stable inorganic nitroxide known as Fremy's radical **(11)** for the oxidation of phenols to (usually *para-*) benzoquinones. Two equivalents of oxidant are required. The first generates a phenoxyl radical which then accepts an oxygen atom from the second. Higher molecular weight phenols, which are relatively insoluble in the aqueous solvent, can usually be oxidised with the aid of a phase-transfer catalyst.

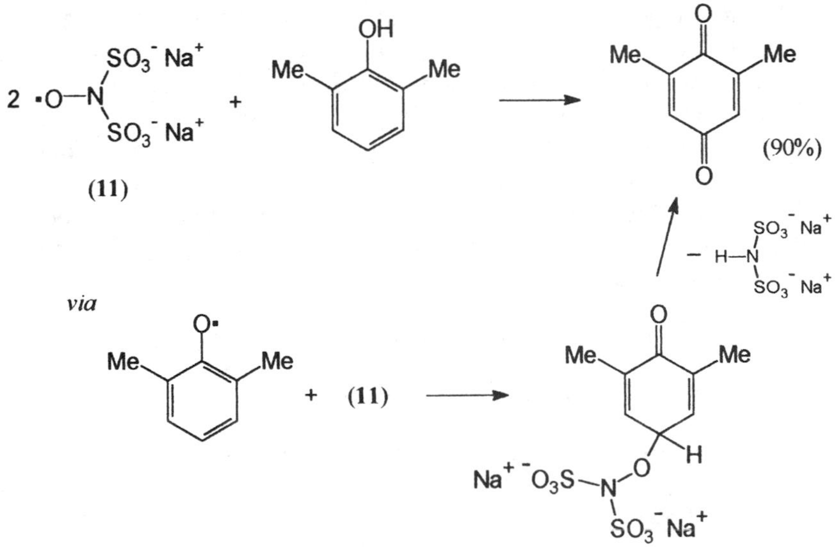

$2\ {\cdot}O{-}N(SO_3^-\,Na^+)_2$ **(11)** + 2,6-dimethylphenol ⟶ 2,6-dimethyl-1,4-benzoquinone (90%)

via

2,6-dimethylphenoxyl radical + **(11)** ⟶ 4-H, 4-[$O{-}N(SO_3^-\,Na^+)_2$]-2,6-dimethylcyclohexa-2,5-dienone ⟶ quinone $-\ H{-}N(SO_3^-\,Na^+)_2$

Alternatively, benzoyl t-butyl nitroxide **(12)**, which is readily soluble in organic solvents, affords a serviceable alternative (the better known di-t-alkyl nitroxides, mentioned in Chapter 5, are insufficiently reactive). Compound **(12)** and related

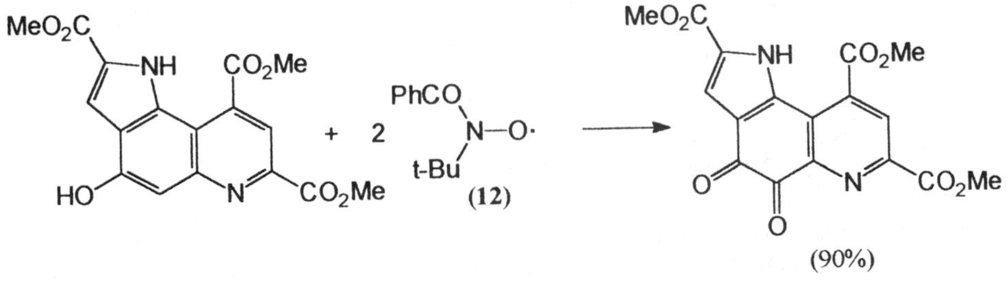

[Reaction scheme: hydroxy pyrroloquinoline trimethyl ester (HO, NH, MeO_2C, CO_2Me, N, CO_2Me) + 2 PhCO(t-Bu)N–O· **(12)** ⟶ corresponding o-quinone (O, O) (90%)]

1 cal = 4.184 J

radicals have also been employed to oxidise allylic and benzylic alcohols [e.g. retinol (**13**)].

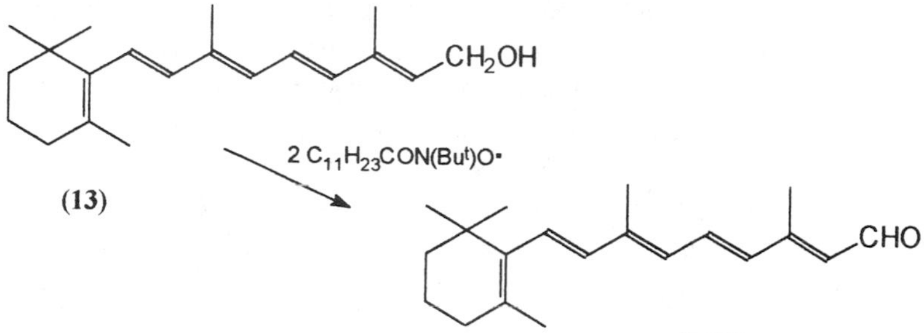

CH_2OH

$2\ C_{11}H_{23}CON(Bu^t)O\cdot$

(**13**)

CHO

(90%)

Phenol oxidations with a variety of other inorganic reagents are also formulated as proceeding *via* phenoxyl radical intermediates, e.g. Schemes 6.15 and 6.16. Similar processes may be key steps in the biosynthesis of numerous alkaloid species,

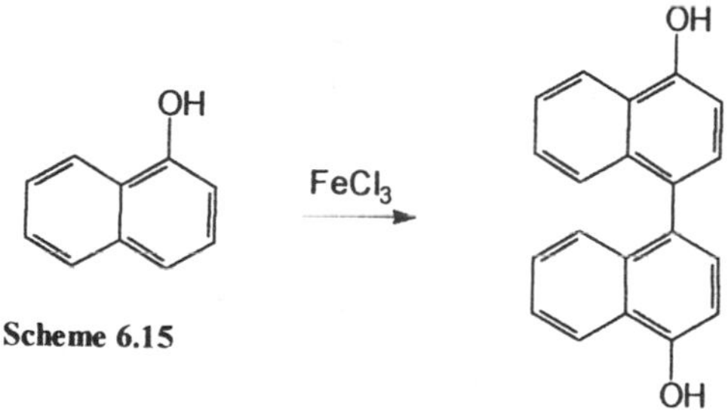

OH FeCl$_3$ OH OH

Scheme 6.15

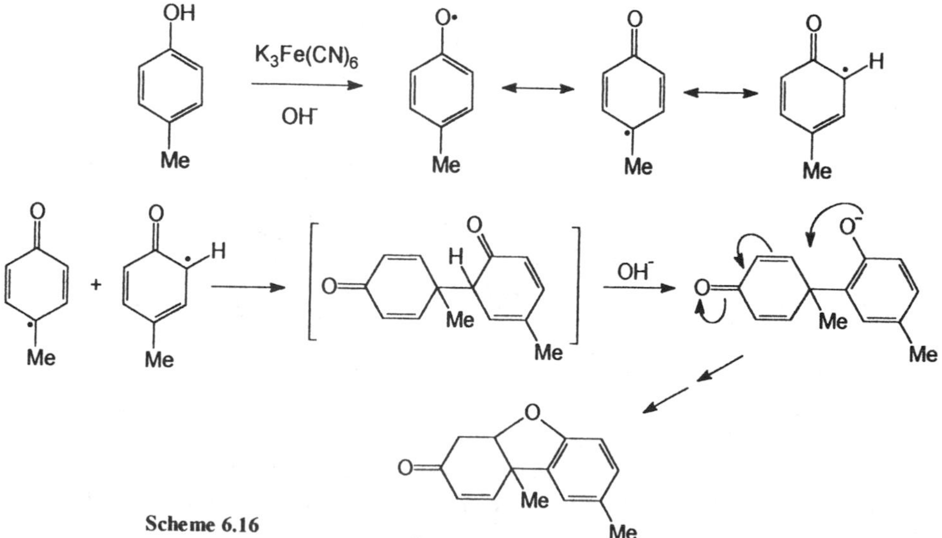

OH $K_3Fe(CN)_6$ OH^- Me O• Me O Me O H Me

O Me + O H Me → [O H Me Me] OH^- → O O^- Me Me

O O Me Me

Scheme 6.16

("Pummerer's ketone")

and several of these have been reproduced *in vitro*, albeit usually in rather poor yield [e.g. the morphinandienone shown in Equation (4)]. Some of the best preparative results with related intramolecular oxidative coupling have been obtained using vanadyl chloride [Equation (5)]. High-yield intramolecular coupling in electrochemical oxidations of related compounds *in which both phenolic oxygens are alkylated* [e.g. Equation (6)] may occur at the radical cation or the dication stage. It is possible that strongly coordinated vanadium in Equation (5) functions rather like the alkyl groups and the reaction follows a similar path. Other mechanistic suggestions for the simple phenol oxidations have included coupling of phenoxyl radical with phenoxide ion.

MeO, HO, NMe, MeO, OH —MnO_2→ MeO, HO, NMe, MeO, O (0.01%) (4)

OH, HO —$VOCl_3$, Et_2O/-78°, then reflux→ O, HO (76%) (5)

OMe, MeO, NMe, MeO, OMe —e^- (Pt anode), CH_3CN/H^+→ OMe, MeO, NMe, MeO, O (52%) (6)

Synthetic applications of coupling of alkyl radicals, e.g. in the Kolbe reaction, have already been noted. A more elaborate example is to be found in the

1 cal = 4.184 J

surprisingly efficient synthesis of cyclodecane (ca. 50%) by thermolysis of the bis-peroxide (**14**) (which is readily prepared from cyclohexanone). In the mechanism for this, outlined in Scheme 6.17, intramolecular coupling occurs twice.

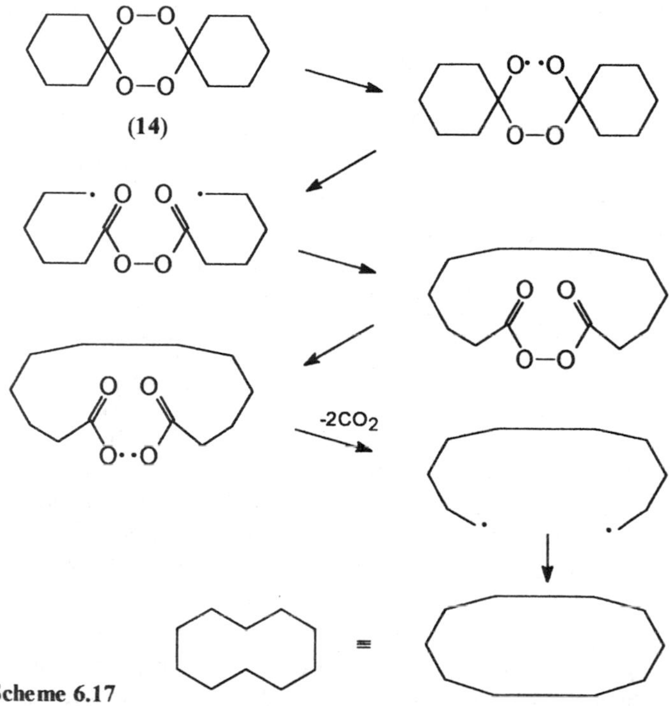

Scheme 6.17

Although the synthesis of a relative of lysergic acid mentioned earlier apparently involved a chain reaction mediated by an aryl radical, aryl radicals have most commonly been produced in non-chain processes, including aroyl peroxide decomposition (see Chapter 9), aryl iodide photolysis (Chapter 3), and diazonium salt reduction (Chapter 3). These processes, especially the last two, have been adapted to intramolecular arylation of adjacent aromatic rings, often with good yields, as in the example given in Scheme 3.6.

These aryl-radical cyclisations, which occur in solvents which might be expected to react rapidly with aryl radicals (e.g. benzene or acetone) owe their success, at least in part, to the kinetic advantage of intramolecularity discussed in Chapter 4. It is interesting to note that this advantage, which frequently amounts to four orders of magnitude or more, is dramatically reduced when intra- and inter-molecular *hydrogen transfers* are compared (down to values usually in the range of ca. 1–50). Nevertheless, as we have already seen, this is still quite sufficient for many intramolecular atom-transfers to be synthetically useful. The quantitative estimation of the kinetic advantage attributable to intramolecularity has been examined in great detail for ionic reactions, some of which serve as models for the intra-complex reactions of enzyme catalysis. It is as yet relatively uncharted territory in radical chemistry.

In another non-chain process, oxidation of diethyl malonate using Mn(III) acetate generates bisethoxycarbonylmethyl radicals **(15)**. These, having two electron-

$CH_2(CO_2Et)_2$ — Mn(III) acetate, HOAc → $\cdot CH(CO_2Et)_2$ **(15)** (Electrophile)

$CH_2(CO_2Et)_2$ — EtO^- → $^-CH(CO_2Et)_2$ (Nucleophile)

withdrawing substituents, behave as electrophilic alkyl radicals, an interesting reversal of the familiar reactivity pattern ("umpolung") in which a nucleophilic (carbanion) centre is formed from from malonate. The manganese reagent yields radicals from many species having relatively acidic C–H bonds. This is exemplified in the efficient synthesis of the bicyclic lactone given at the end of Chapter 1. Nickel peroxide (NiO_2) is another reagent which seems able to generate radical centres by oxidising molecules containing relatively acidic hydrogen atoms.

Umpolung in the reverse sense is illustrated by α-alkoxyalkyl radicals which are nucleophilic, whilst the corresponding ion is the electrophilic species $R'CH{=}O^+R$.

Yet another non-chain source of carbon-centred radicals uses organocobalt compounds, and was originally patterned on the chemistry of coenzyme B_{12}.[1] Thus it has been found that certain Co(I)-complexes [generated *in situ* by reduction of Co(II) or Co(III) precursors] react with organic halogen compounds to give organocobalt(III) species with a carbon–cobalt bond which cleaves homolytically under mild thermal or photochemical conditions. This is exemplified in Scheme 6.18.[2] The key intermediate **(18)** in the scheme may be generated directly from the dihydrobenzofuran **(16)**, or from the aryl iodide **(17a)** following cyclisation of the derived aryl radical **(17b)**.

In a further example of this type of process the organocobalt derivative **(19)**, prepared from the corresponding acid chloride, dissociates on heating, and the resulting radical cyclises to give a β-lactam, transformable into the antibiotic

[1]See Chapter 10, p. 143.

[2]The ligands employed are more electrophilic than B_{12} or those in the familiar cobaloximes. Similar experiments with these were unsuccessful. However, interesting radical chemistry has been developed with allyl- and benzyl-cobaloximes.

1 cal = 4.184 J

thienamycin. This 4-*exo* cyclisation is unusual. Evidently, under favourable circumstances, it can afford a useful source of four-membered rings.

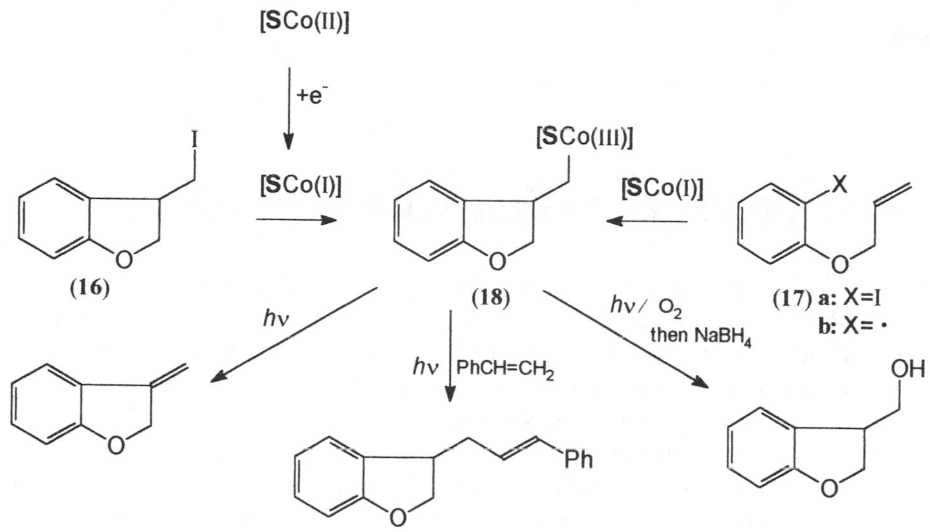
[SCo(II)]
+e⁻
[SCo(I)]
[SCo(III)]
[SCo(I)]
I
X
(16)
(18)
(17) a: X=I
b: X= •
hν
hν PhCH=CH2
hν/ O2 then NaBH4
Ph
OH

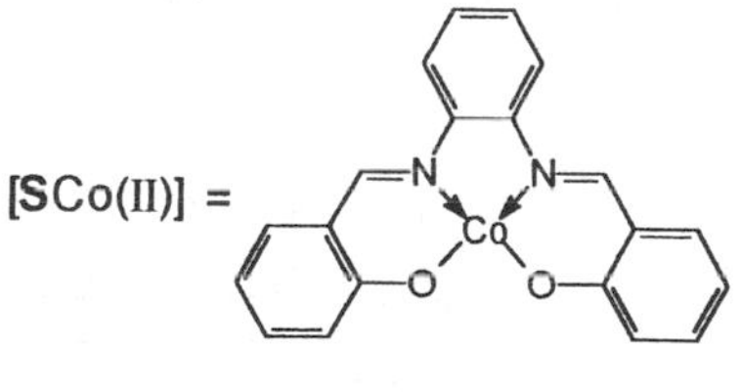
[SCo(II)] =
N
N
Co
O
O

Scheme 6.18

"Cobalt(II)salophen"

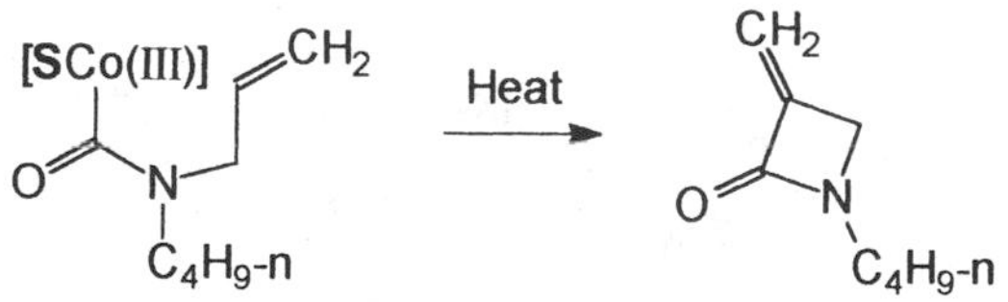
[SCo(III)]
CH2
O
N
C4H9-n
Heat
CH2
O
N
C4H9-n

(19)

7

Some stereochemical considerations

In the context of synthetic strategy, the facts that alkyl radicals are generally planar at the radical centre, and that vinyl radicals, although generally bent at the radical centre, undergo rapid stereoinversion, might seem to indicate that formation of new bonds at these centres is always going to result in mixtures of stereoisomers [e.g. Equations (1) and (2)]. Early observations, such as the near 50:50 mixtures of diastereomers from dimerisation of alkyl radicals of the general form XYZC·,[1] or from radical addition of XY to RCH=CHR, seemed to support this generalisation. There were few exceptions, although additions of HBr or Br_2 to alkenes do show a marked preference for *trans*-addition, which complements other evidence that there

(1)

(2)

[1]Interesting exceptions are found where radical pairs are formed in a low-temperature glassy matrix, e.g. by azo-compound photolysis, and collapse of the pair then forms product in that constrained environment. Thus ultraviolet irradiation of a solid solution of the *meso* isomer of PhCMe(i-Pr)–N=N–CMe(i-Pr)Ph in a hydrocarbon glass gives dimer [PhCMe(i-Pr)–CMe(i-Pr)Ph] which also is exclusively in the *meso* form.

1 cal = 4.184 J

is some (unsymmetrical) bridging by bromine in the intermediate β-bromoalkyl radicals (**1**).

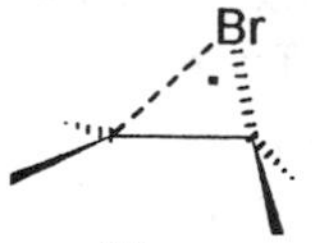

(**1**)

Another circumstance in which considerable diastereoselectivity has been observed in reactions of chirally substituted alkyl radicals is that in which both an asymmetric carbon centre and a carboxylic ester group are attached to the radical centre. This is exemplified in Equation (3), in which the product isomer shown predominates over its diastereomer by at least 30:1. The result has been interpreted in terms of an approximately planar oxyallyl radical (**2**) reacting from a preferred conformation, (**3**). In this, the steric bulk of phenyl is considerably greater than that of methoxyl, so that the tin hydride reagent delivers hydrogen to the less hindered upper face. Although by carrying out, for example, allylation of such alkyl radicals at low temperatures very high selectivities have been achieved, the stereochemical arguments do not appear to be entirely corroborated by the results of related experiments.[1]

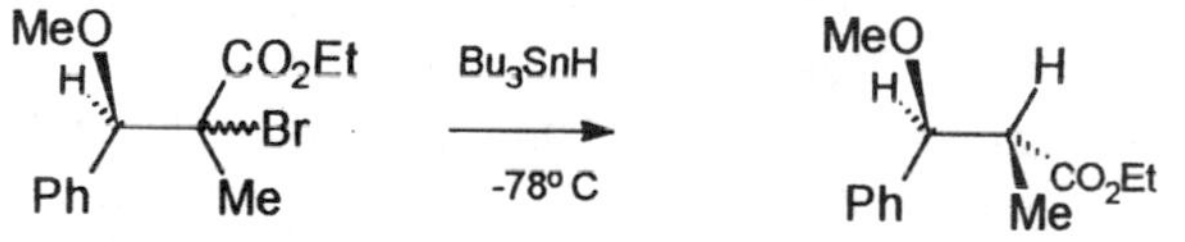

MeO, H, Ph, CO_2Et, Br, Me —(Bu_3SnH, -78° C)→ MeO, H, Ph, H, CO_2Et, Me (3)

(diastereoselectivity > 30:1)

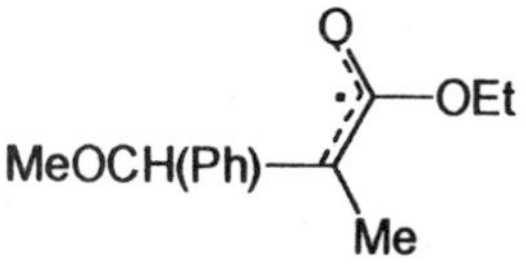

MeOCH(Ph)—, O, OEt, Me (**2**)

MeO, H, Ph, H—$SnBu_3$, O, OEt, Me (**3**)

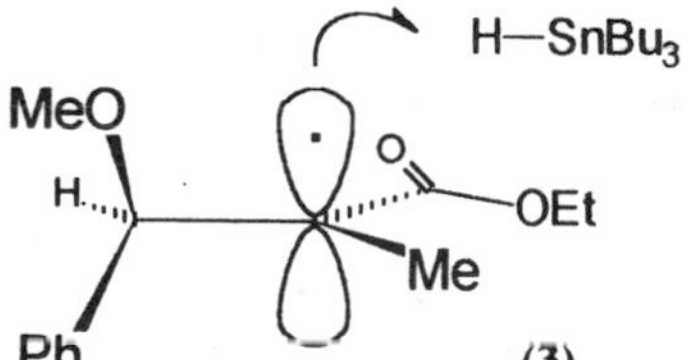

Where, in more obviously rigid systems, steric effects introduce significant geometrical constraints upon the course of a reaction, quite large diastereo-selectivities have also come to light. For instance, the 2-methoxycyclopentyl radical (**4**) adds to acrylonitrile such that the preference for a *trans* relationship between the

[1]The interpretation given is in terms of the conformation of the alkyl radical shown in (**3**). The reader is reminded of the Curtin–Hammett principle, which points out that competition of this kind should properly be discussed in terms of the relative stabilities of the *transition states* leading to the two diastereomeric products.

methoxy and cyanoethyl groups is ca. 5:1. In a similar manner, addition of "t-Bu–H" to methylmaleic anhydride by the organomercury–sodium borohydride route (Chapter 6) gives the *cis*-dialkylsuccinic anhydride (**5**) in 88% yield. Less than 5% of the *trans*-isomer is obtained. In this case the intermediate mercuric hydride delivers hydrogen *trans* to the bulky t-butyl substituent.

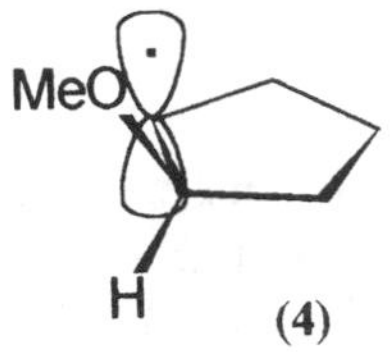

MeO

H

(**4**)

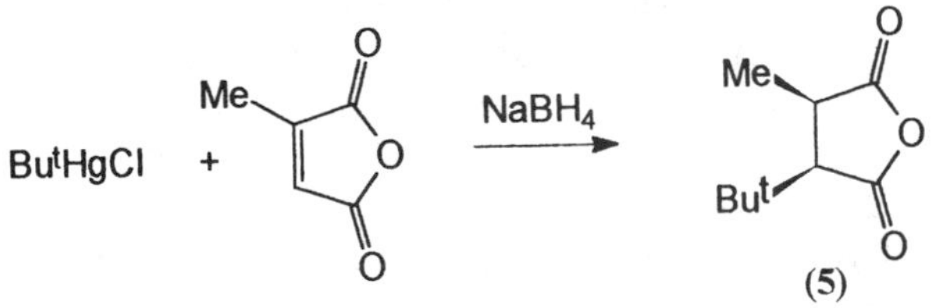

Bu^tHgCl + Me ... $\xrightarrow{NaBH_4}$ Me ... Bu^t

(**5**)

Within the last few years methods have begun to be developed whereby this type of effect can be incorporated into the planning of synthetic strategy. Where the diastereomeric control can be effected by a removable chiral auxiliary group, then effective enantiocontrol can be achieved. One early example of this was in a synthesis of muscone (**6**), outlined in Scheme 7.1. In this, radical cyclisation was by addition to a β-acylacrylamide moiety in which the nitrogen atom formed part of an (enantiomerically pure) *trans*-2,5-dimethylpyrrolidine. Four distinct products were isolated from the initial cyclisation. *endo*-Addition predominated (8:1) and

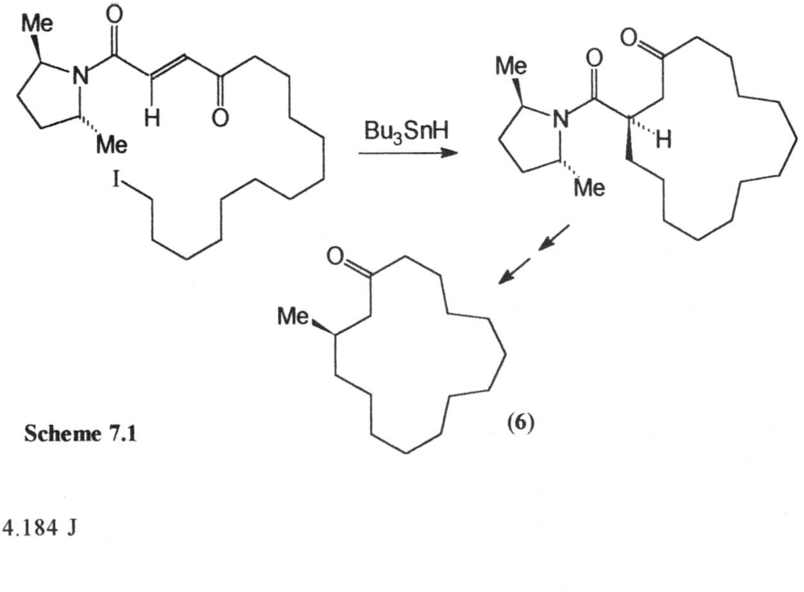

$\xrightarrow{Bu_3SnH}$

(**6**)

Scheme 7.1

1 cal = 4.184 J

showed a very large diastereoselectivity (14:1). In contrast diastereoselection in formation of the *exo*-adduct was small. In this example, although the product of the cyclisation was carried through to muscone, the hydrolysis of the amide proved difficult. There are, however, indications that more tractable chiral auxiliaries may soon be available for similar reactions.

The effect of the chiral pyrrolidine in this synthesis of muscone can be interpreted as indicated in structure (**7**). The whole acrylamide unit is essentially coplanar so that one face of the α-carbon is very much more accessible than the other (irrespective of which amide rotamer is involved, since a 180° rotation restores the original structure).

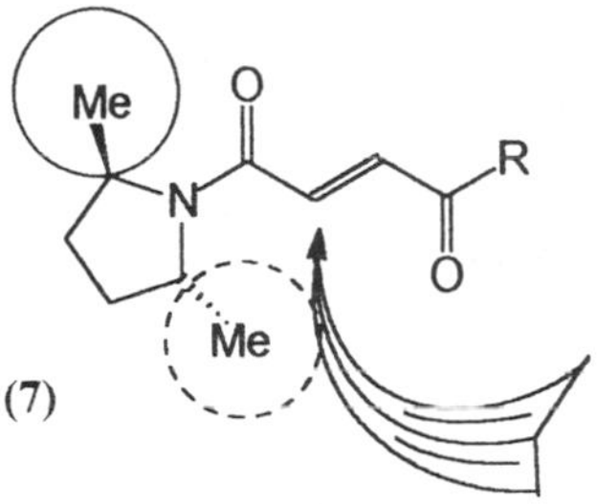

(7)

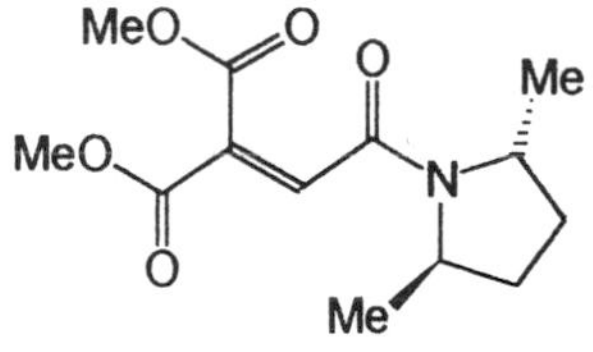

(**8**)

The above strategy has also been successful in intermolecular additions, for example to compound (**8**), but once again high diastereoselectivity is achieved only when addition is to the carbon α to the amide function. Selectivity in attack at the β-position has, however, been accomplished in more recent experiments. This promises much greater versatility, but the chiral auxiliary used is structurally complex and requires significant synthetic effort for its preparation. The auxiliary is constructed from Kemp's triacid (**9**), and its success depends on adoption by the alkene derivative of the conformation shown in (**10**). In the example given, this will clearly be favoured by repulsion between the carbonyl dipoles of the imide, and by the absence of any significant steric interference if the acyclic carbonyl and the C–C double bond adopt the *s-cis* conformation. Addition of t-butyl radicals was even more selective than expected, in that whilst diastereoselectivity α to the ester group

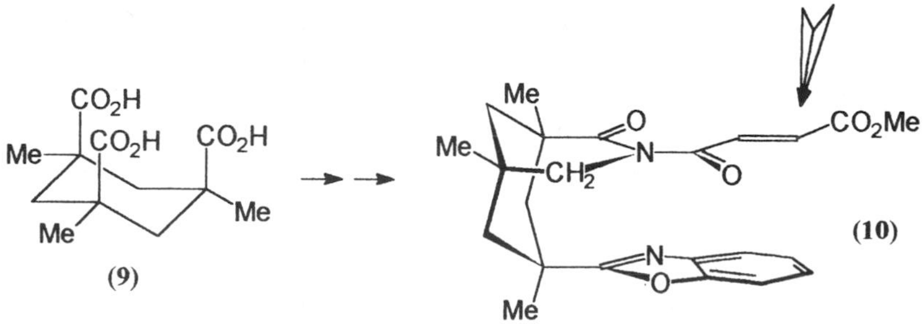

(9)

(**10**)

was (at −40°C) 97:3 in the expected direction, *no attack α to the imide was detected!*

Some interesting examples of asymmetric induction in intramolecular radical additions have been demonstrated in manganese(III) acetate oxidations of olefinic acetoacetates. For example, the major product (50%) from oxidation of the (+)-phenylmenthyl keto-ester (**11**) is a 1:7 mixture of the diastereomers (**12a**) and (**12b**),[1] from which the more abundant isomer was transformed into *O*-methylpodocarpic acid (**13**) having the natural configuration.

OMe Me OMe Me H O Me X O H Me O X (11) Mn(III) acetate (12a) + OMe OMe Me Me O Me H X O (12b) Me H O OH (13) X = O Ph

A completely different approach to chirality transfer in radical chemistry involves hydrogen abstraction from a racemic substrate by an enantiomerically pure chiral radical. Different reaction rates would be expected for reaction with the two substrate enantiomers, since the transition states will be diastereomeric (Scheme 7.2). This should manifest itself in partial resolution of any unreacted substrate. The principle was first demonstrated in the 1970s, but enantiomeric excesses achieved were small. Since the radical centre in the atom-abstracting species (**14**) will generally not be the same as the chiral centre, the poor selectivity is unsurprising. However, in recent work, with more careful matching of the steric demands of abstracting radical and substrate, some quite large enantioselectivities have been

[1]Note that the isomers (**12**) result from 6-*endo*, rather than 5-*exo* cyclisation. This reflects reversibility of the intramolecular addition of the doubly carbonyl-stabilised radical (see Chapter 6); i.e. product formation here is under thermodynamic control.

1 cal = 4.184 J

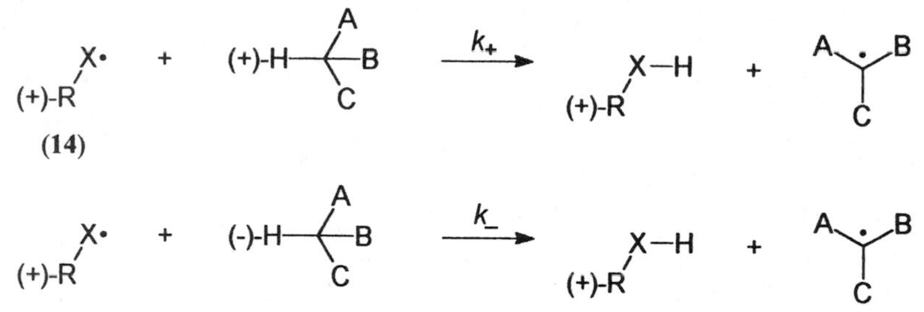

(+)-R–X• (14) + (+)-H–C(A)(B)(C) $\xrightarrow{k_+}$ (+)-R–X–H + A–Ċ(B)–C

(+)-R–X• + (-)-H–C(A)(B)(C) $\xrightarrow{k_-}$ (+)-R–X–H + A–Ċ(B)–C

Scheme 7.2 ($k_+ \neq k_-$)

observed. The example illustrated below, in which one enantiomer of substrate **(16)** is removed very much faster than the other, has the further advantage that the reaction is catalytic in chiral reagent **(15)**. It represents a further instance of polarity reversal catalysis (see Chapter 6, p. 89) involving, as it does, a highly electropositive amine-boryl radical.

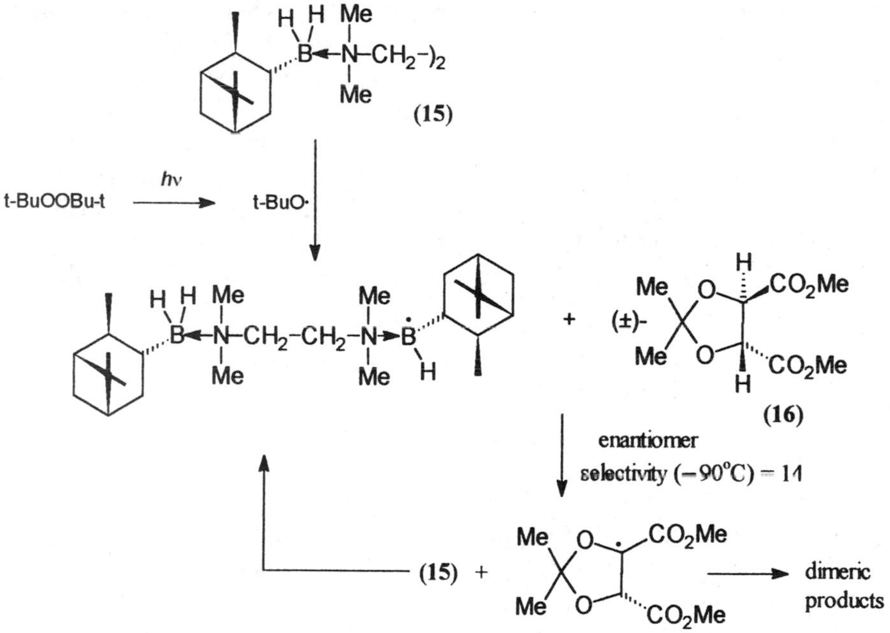

H H Me

B←N–CH$_2$–)$_2$

Me

(15)

t-BuOOBu-t $\xrightarrow{h\nu}$ t-BuO·

H H Me Me

B←N–CH$_2$–CH$_2$–N→Ḃ

Me Me H

\+ (±)- Me Me O O H H CO$_2$Me CO$_2$Me

(16)

enantiomer selectivity (−90°C) = 14

(15) + Me Me O O CO$_2$Me CO$_2$Me ⟶ dimeric products

8

Radical ions and electron transfer

Single electron transfer (SET), involving the formation or reaction of radical ions, is surprisingly commonplace in organic chemistry. However, like the electrically neutral radicals that have been the principal topics of discussion in this text, radical ions most commonly occur only as transients. One well-known exception is encountered in the familiar blue colour which is indicative of the absence of both moisture and oxygen when ether solvents are dried by refluxing over sodium in an inert atmosphere and in the presence of benzophenone. When all moisture has been removed, the coloured benzophenone ketyl (**1**) persists in the solution. Like this ketyl, many radical ions are sensitive to oxygen and moisture (or other protic compound), but, in the absence of these, are quite stable. Consequently, spectroscopic studies of radical ions proved to be a fertile area of early ESR researches; one example is the radical anion of naphthalene, whose spectrum was illustrated in Chapter 5. This species may be observed as the yellowish colour which develops when a dry THF solution of naphthalene is brought into contact with a clean potassium mirror deposited *in vacuo* on the walls of a suitable vessel.

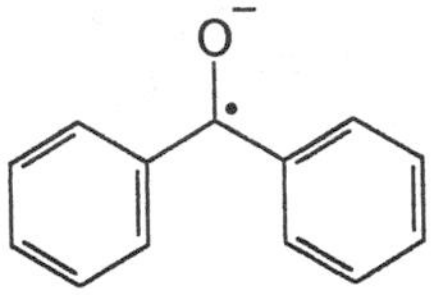

O^-

(**1**)

Modern electrochemical techniques have proved invaluable in the study of radical ions and of the reactions in which they participate. For example, these methods can usually give quantitative information on the ease of one-electron oxidation or reduction of various organic substrates. One important, if unsurprising, result to emerge from such studies is the sensitivity of redox potentials to solvent. This is reflected in a much stronger solvent-dependence of reactions involving radical ions than is encountered with reactions mediated exclusively by neutral

1 cal = 4.184 J

radicals (an effect which is carried over into some examples of molecule-induced homolysis – Chapter 3). Electrochemistry also provides a technique for ESR observation of radical ions using an experimental set-up in which an appropriate precursor is anodically oxidised or cathodically reduced within a specially designed electrolysis cell inserted in the microwave cavity of the spectrometer. This procedure is probably the most effective general method for the production of a variety of radical anions for spectroscopic study, notably those of unsaturated carbonyl compounds (including quinones), aromatic hydrocarbons, nitroaromatics, e.g. (**2**), and many others. Common solvents are acetonitrile, DMSO, DMF, etc.

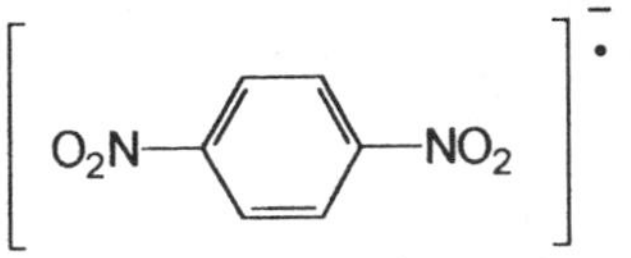

$$[O_2N-C_6H_4-NO_2]^{\bullet -}$$

(**2**)

Historically, the first observed (albeit unrecognised, for the year was 1836) radical ion was the benzil semidione (**4**), responsible for the blue colour which may be detected when benzil is treated with potassium hydroxide. The colour is more readily observed if benzoin is also present, when formation of the semidione is due to "dismutation" of the dianion (**3**) and benzil.

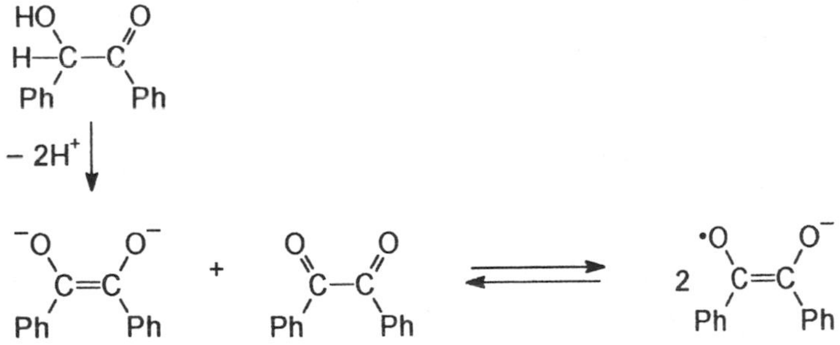

$$PhCH(OH)C(=O)Ph \xrightarrow{-2H^+} {}^{-}O(Ph)C{=}C(Ph)O^{-} + PhC(=O)C(=O)Ph \rightleftharpoons 2\ {}^{\bullet}O(Ph)C{=}C(Ph)O^{-}$$

(**3**) (**4**)

Many of the more stable radical ions are characterised by the presence of aryl substituents, or other means of extensive electron delocalisation. For example, radical cation salts of some triarylamines, e.g. (**5**), can be obtained in crystalline

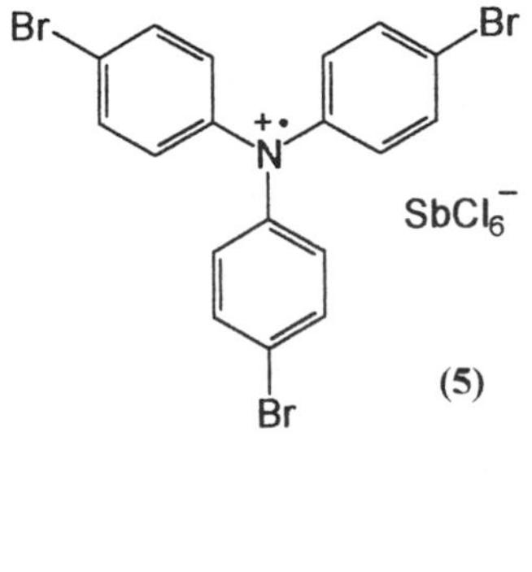

$$(4\text{-}BrC_6H_4)_3N^{+\bullet}\ SbCl_6^{-}$$

(**5**)

form (provided all three *para*-positions are blocked by substitution), and have found applications as one-electron oxidising agents. Unusually, one of the most widely known radical cations, (**7**), familiar because it is is a critical intermediate in the herbicide action of the bipyridyl derivative "paraquat" (**6**), is generated from (**6**) by one-electron *reduction*.

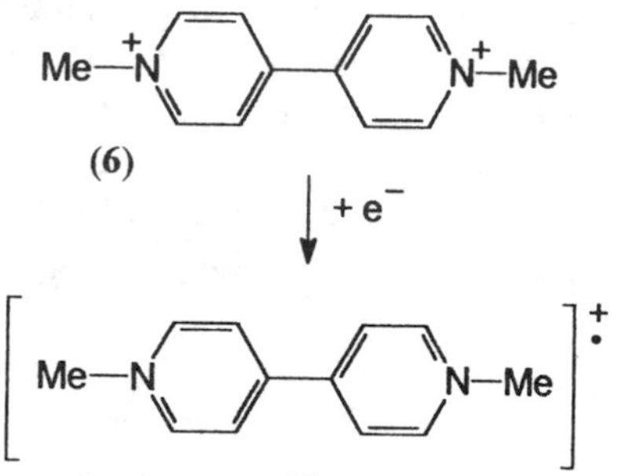

Me—N⁺ ... N⁺—Me

(**6**)

+ e⁻

[Me—N ... N—Me]$^{+\bullet}$

(**7**)

The reader is likely to be familiar with the transient radical cations that are formed by electron bombardment of target molecules in the mass spectrometer. Less familiar is the production of similar species, sufficiently long-lived for ESR study, by the interaction with ionising radiation of organic substrates frozen into a suitable glassy matrix[1] at low temperatures (usually liquid N_2 and below). This procedure has also been used extensively for spectroscopic study of radical cations, as well as neutral radicals which may be formed by radical ion dissociation in the matrix. Provided these species cannot react with molecules of the matrix they are indefinitely long-lived, since diffusion through the matrix and bimolecular decay, e.g. by dimerisation, is strongly inhibited if not impossible. Two examples are shown in Scheme 8.1. Whilst this affords a means of studying many radicals and radical ions at leisure, there is the severe limitation that, since the radicals are held rigidly in the matrix, they are not tumbling, so that anisotropic effects are not averaged and if the radicals have been trapped in a glassy matrix lacking any structural order their spectra will comprise a summation of spectra of radicals in all possible orientations with respect to the applied magnetic field. The resulting spectra are frequently very broad and far from straightforward to interpret. A spectrum obtained in this way is traced in Fig. **8.1**. (For further discussion see Section 10.3).

[1]Various single component and mixed solvent matrices have been used. The choice is critically dependent on the phenomena to be studied. For example, electrons will migrate in some matrices but not in others. 2-Methyltetrahydrofuran and 3-methylpentane are common examples.

1 cal = 4.184 J

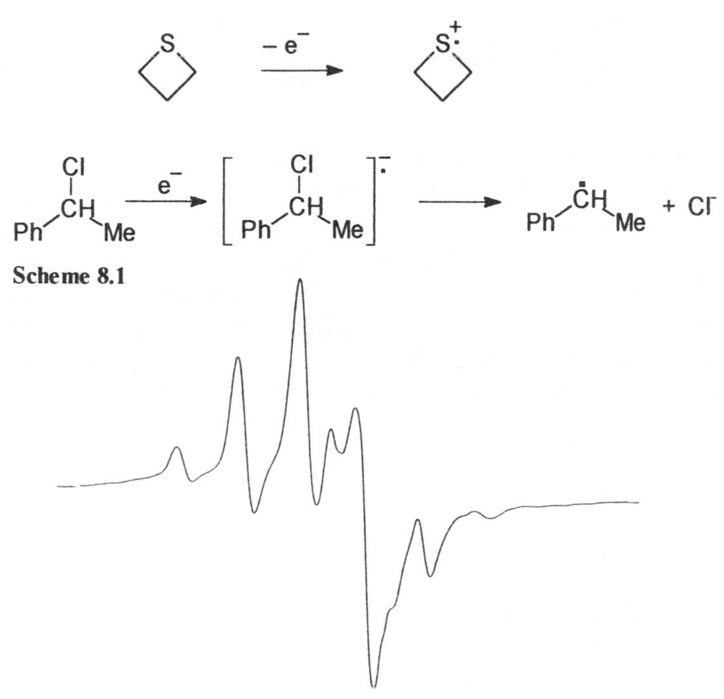

Scheme 8.1

Fig. 8.1: The ESR spectrum of the thietane radical cation (see Scheme 8.1), trapped in a $CFCl_3$ glass at 115 K. (Reproduced with permission from F. Williams, *J. Amer. Chem. Soc.*, **109**, 6778 (1987)).

A different type of matrix-isolation technique, in which rotational motion of the trapped radicals is much less restricted, uses adamantane as the matrix medium. We shall return to this in the next chapter.

Electronic structure: Whereas radicals are generally categorised as being of σ- or π-type, for radical ions there are three categories. Whilst the structures of those referred to as being of π-type are reasonably self-evident, for example the radical anion and radical cation of naphthalene, in which the unpaired electrons reside in the lowest energy antibonding π-orbital (π*-orbital) and the highest energy π-bonding orbital of the hydrocarbon respectively, σ-radical ions are typified by species which have lost an electron from a σ bonding orbital, as might occur from a saturated hydrocarbon in the mass spectrometer. The third category, referred to as *p*-radical ions (sometimes "n" for non-bonding), are typified by aliphatic amine or ether radical cations where there is a single electron in a non-bonding orbital on the heteroatom.

An interesting variation on the theme of hypervalent radicals, such as phosphoranyl ($R_4P\cdot$ – see Chapter 2, p.14) and the related sulphuranyl ($R_3S\cdot$) radicals, has been encountered during one-electron oxidations, in particular of sulphides. This variant arises by association between the electron-deficient radical cation and the lone-pair of an unoxidised molecule: e.g. $R_2S^{\cdot+}$ + $:SR_2$ → $R_2S^{+}\text{–}S(\cdot)R_2$. The resulting dimeric radical cation is commonly represented as having a three-electron bond, $[R_2S \therefore SR_2]^{+}$.

Some reactions involving radical ions: Aqueous hydrolysis of naphthalene radical anion gives a mixture of naphthalene and dihydronaphthalenes. Benzophenone ketyl similarly gives benzhydrol and benzophenone. The radical anions of naphthalene, benzophenone, and nitrobenzene all reduce molecular oxygen to superoxide, $O_2^{\cdot-}$ by SET. Alkyl iodides capture the electron from naphthalene radical anion and subsequently dissociate to iodide and an alkyl radical. A complicated series of reactions ensues, and the products include alkylated dihydronaphthalenes, most probably resulting from a pathway such as that shown in Scheme 8.2.

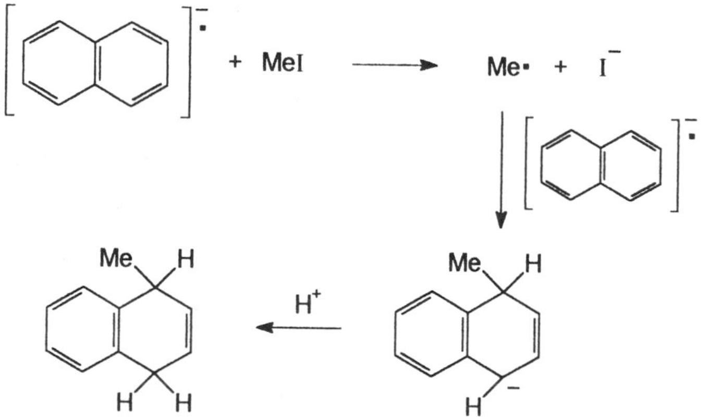

Scheme 8.2

One example of the practical utility of (**5**), is the near quantitative oxidation of *N*-vinylcarbazole in methanol to form the dimeric di-ether (**8**). This example demonstrates that pre-prepared radical ions can be useful oxidising or reducing agents. More usually, however, formation of transient radical ion intermediates involves inorganic reagents, or may occur at an electrode. Obvious one-electron reducing agents which will be encountered in most general organic chemistry texts include metals such as magnesium employed in the pinacol reaction, as well as the more recently introduced samarium iodide. These and related systems were mentioned in Chapter 6. Another familiar example is the preparative reduction of naphthalene to 1,4-dihydronaphthalene by sodium under Birch conditions; the naphthalene radical anion is undoubtedly an intermediate. Chromous salts have also been used for reduction of a variety of functional groups, especially halides (→ RH), but including epoxides, nitro-compounds and others.

1 cal = 4.184 J

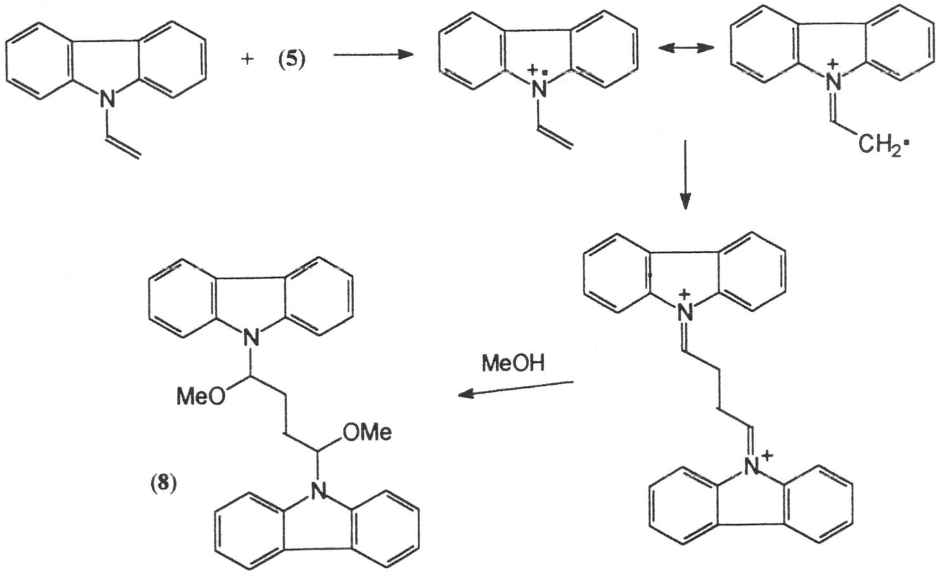

+ **(5)**

MeOH

MeO

OMe

(8)

Common inorganic one-electron oxidising agents include alkaline ferricyanide [hexacyano-ferrate, $(Fe(CN)_6^-$; e.g. phenol oxidation, Chapter 6)] and potassium persulphate ($K_2S_2O_8$).

In the same way that long-lived or transient radical anions may be captured by electrophiles (e.g. H^+), radical cations may be intercepted by nucleophiles. We have noted earlier how free aliphatic carboxylate (acyloxyl; $RCO_2\cdot$) radicals generally lose CO_2 far too rapidly to be trapped. Therefore the efficient formation of α-naphthyl acetate at the anode when a solution containing both sodium acetate and naphthalene is electrolysed cannot be interpreted in terms of acetoxyl radical attack on naphthalene. Instead it involves oxidation of naphthalene to the radical cation which is subsequently intercepted by acetate (Scheme 8.3).

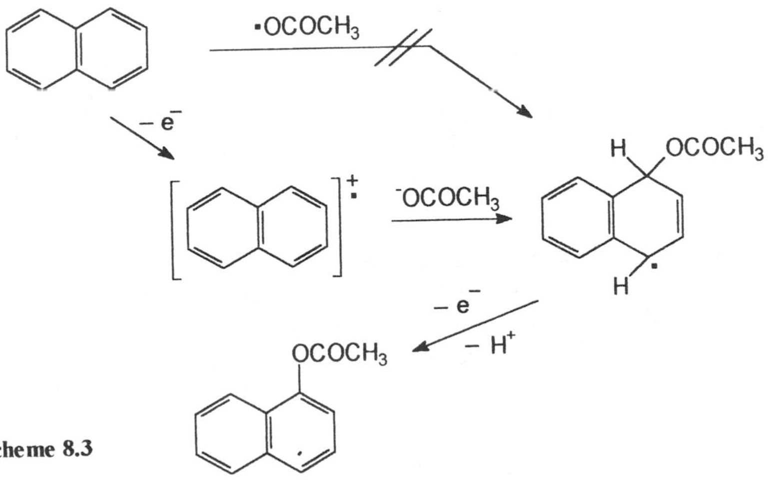

•OCOCH3

– e⁻

⁻OCOCH3

H OCOCH3

H

– e⁻

– H⁺

OCOCH3

Scheme 8.3

From time to time efforts have been made to reinterpret heterolytic substitution reactions in terms of radicals and electron transfer. Whilst it seems that this can generally be discounted, there are some interesting cases in which radicals and radical ions are unquestionably involved in both aliphatic and aromatic substitutions. One of the best-known examples of this is found in the reactions of tertiary nitroalkanes (usually incorporating an electron-withdrawing substituent α to the nitro group). These undergo efficient displacement of the nitro-group by a variety of nucleophiles, notably stabilised carbanions. An example is shown in Equation (1), and the mechanism for this is outlined in Scheme 8.4. Reactions of this kind have been designated $S_{RN}1$, involving as they do unimolecular scission of $RX^{\cdot-}$,

$$\mathrm{Me_2\bar{C}{-}NO_2} + \mathrm{Me_2C(CO_2Et)NO_2} \longrightarrow \underset{95\%}{\mathrm{EtO_2C{-}CMe_2{-}CMe_2{-}NO_2}} + \mathrm{NO_2^-} \qquad (1)$$

Initiation:

$$\mathrm{Me_2C(CO_2Et)NO_2} \xrightarrow{+\,e^-} \mathrm{Me_2C(CO_2Et)NO_2^{\cdot-}}$$

Propagation:

$$\mathrm{Me_2C(CO_2Et)NO_2^{\cdot-}} \longrightarrow \mathrm{Me_2\dot{C}{-}CO_2Et} + \mathrm{NO_2^-}$$

$$\mathrm{Me_2\dot{C}{-}CO_2Et} + \mathrm{Me_2\bar{C}{-}NO_2} \longrightarrow \mathrm{EtO_2C{-}CMe_2{-}CMe_2{-}NO_2^{\cdot-}}$$

$$\mathrm{EtO_2C{-}CMe_2{-}CMe_2{-}NO_2^{\cdot-}} + \mathrm{Me_2C(CO_2Et)NO_2} \longrightarrow \mathrm{Me_2C(CO_2Et)NO_2^{\cdot-}} + \mathrm{EtO_2C{-}CMe_2{-}CMe_2{-}NO_2}$$

Scheme 8.4

reminiscent of ionisation of RX in S_N1 reactions. Several closely related reactions have also proved useful, for instance the formation of alkenes from secondary nitroalkanes exemplified in Scheme 8.5. The bromo-compound is readily available by bromination of nitrocyclohexane, and the subsequent elimination appears to be initiated by electron transfer from the thiophenoxide. Diphenyl disulphide is

1 cal = 4.184 J

produced, but the chief electron donor may be $N_2O_4^{\cdot-}$ formed in a chain-propagation sequence by elimination from the radical anion of the vicinal dinitro-compound. The principal fate of the nitro-groups is as free NO_2. The elimination is accelerated by light, and inhibited by the presence of certain radical scavengers. Whatever the mechanistic detail, many polysubstituted alkenes are efficiently accessed by this procedure.

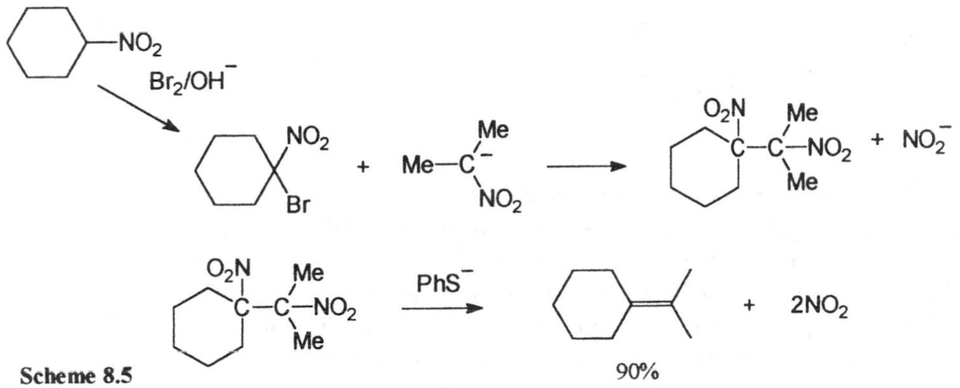

Scheme 8.5

In aromatic chemistry, too, overall nucleophilic substitution can take place *via* radicals and radical anions. These substitutions (typified by the example illustrated in Scheme 8.6) are commonly effected in liquid ammonia, and are usually accelerated by light and inhibited by radical trapping agents.

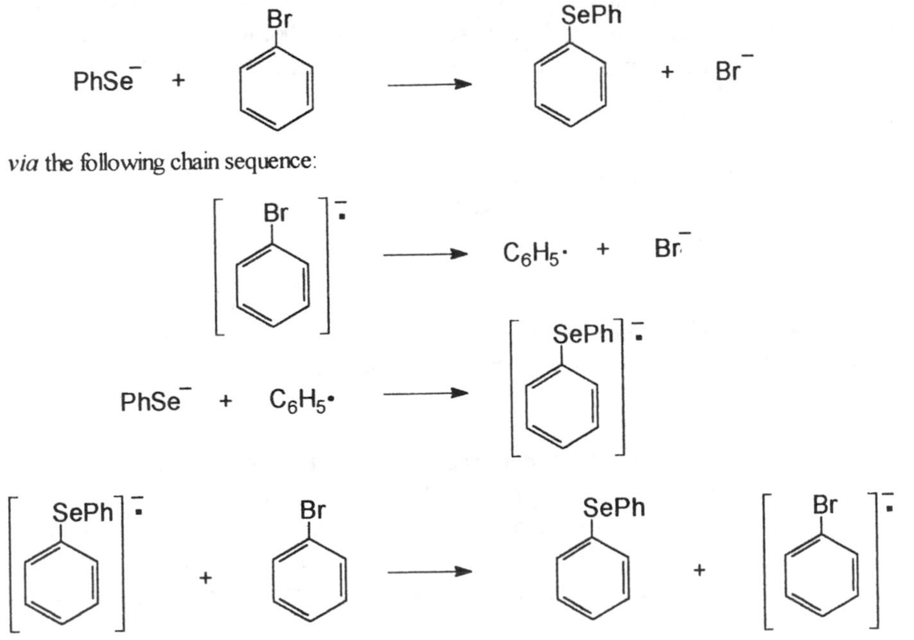

via the following chain sequence:

Scheme 8.6

Of perhaps greater significance for the synthetic chemist, it has been found that, with suitable substrates, excellent yields of substitution product can be obtained when the nucleophile is the enolate of a ketone or ester. In an intramolecular example, the tetralone (**9**) was obtained in 99% yield. Interestingly, stabilised enolates, such as those from β-diketones, are unreactive.

$h\nu$

(**9**)

This chapter would be incomplete without some mention of work on pericyclic processes involving radical ions. Very little of significance appears to be known about this aspect of *neutral* radical chemistry; the case of cyclopropyl ring-opening has been mentioned (Chapter 4), but there is no compelling experimental work on whether the process prefers a conrotatory or a disrotatory path, and some theoretical work suggests that both are "forbidden". By contrast, there is a considerable literature on pericyclic processes involving radical ions.

It is well known that certain pericyclic reactions are catalysed by Lewis acids. In the limit of complete electron transfer is the Diels–Alder dimerisation of, for example, 1,3-cyclohexadiene in the presence of a small quantity of tris-*p*-bromophenylaminium hexachloroantimonate, (**5**); this reaction is rapid at −70°C. In the absence of the catalyst it is slow at 200°C. The assumption is that the diene radical cation behaves as a highly electron-deficient alkene, reacting with a neutral molecule as the diene component. Indeed, orbital symmetry considerations do lead

$(p\text{-BrC}_6H_5)_3N^{\dot{+}}\ SbCl_6^-$ +

(**5**)

$+e^-$

to the prediction that in Diels–Alder reactions one-electron oxidation of the alkene component facilitates an allowed process. On the other hand, it has been concluded that if one-electron oxidation is of the diene component then the cycloaddition should be forbidden. However, in the example of catalysed cycloaddition of (**10**) to (**11**), for which analysis of the experimental data, including the measured

1 cal = 4.184 J

oxidation potentials of the two dienes, shows that essentially only **(10)** must react as cation radical, both **(12)** and two stereoisomers of **(13)** are formed in comparable amounts. Cycloadducts **(13)** result from the "forbidden" pathway.

Catalysis of a number of other reactions by **(5)** has also been shown to be synthetically important. Typically, epoxidation of alkenes by catalysed oxygen-atom transfer from selenium dioxide is not only efficient, but occurs with quite different selectivity from perbenzoic acid oxidations.

Elcctron-transfer reactions are quite different from other bimolecular processes in that any kinetic barrier, which is normally only slightly greater than the heat of reaction, derives from reorganisation of bonding in the reacting partners and the re-positioning of neighbouring solvent molecules.[1]

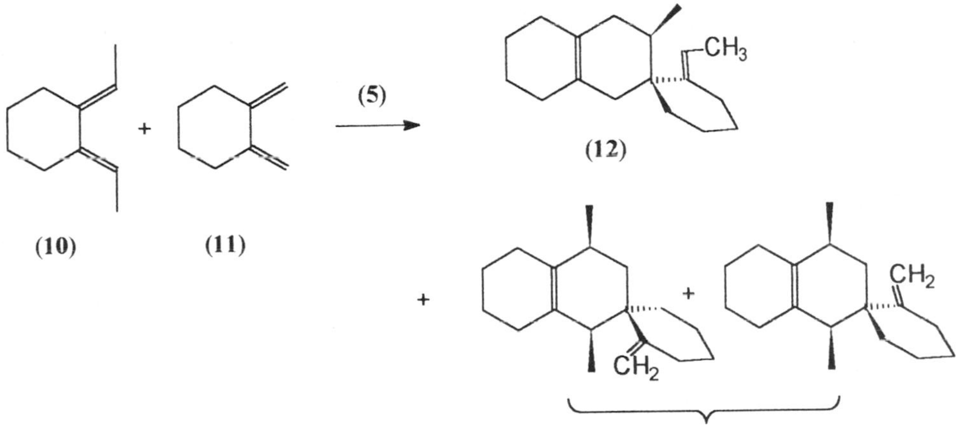

(10) + (11) —(5)→ (12) + (13)

CH3 CH2 CH2

[1]The relevant theoretical treatment was originally devised for electron transfer between inorganic species by R. A. Marcus.

9

Some special topics

9.1 FRONTIER ORBITAL APPROACH TO REACTIVITY

For significantly exothermic reaction steps, patterns of reactivity can often be addressed in terms of interactions between the "frontier" (i.e. highest energy occupied or lowest energy unoccupied) molecular orbitals (HOMO and LUMO respectively) in the two reacting species. As two molecules come together, the

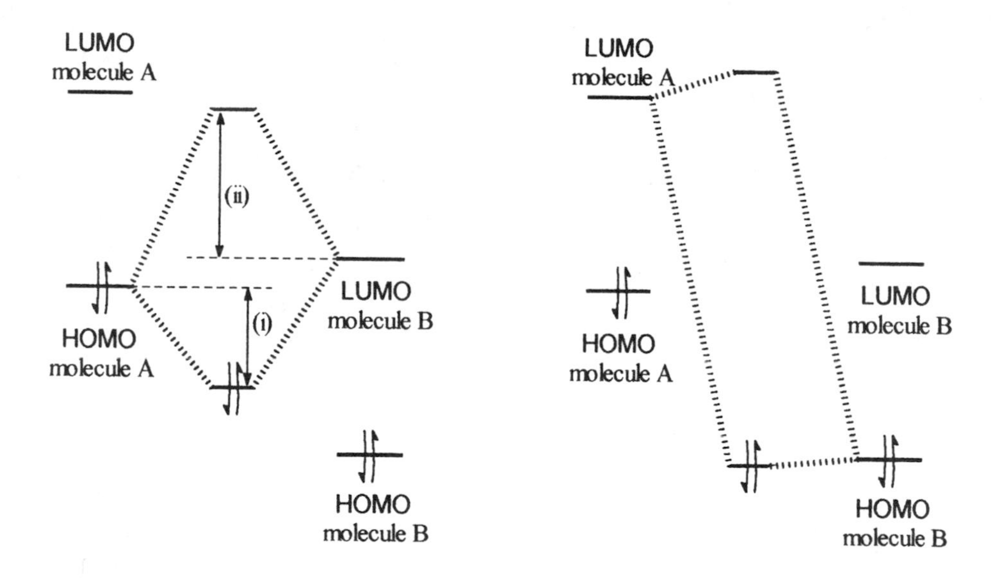

Fig. 9.1: Interaction between frontier orbitals of two molecules A and B. On the left is shown the strong interaction between orbitals of similar energy, on the right is the much smaller effect when there is a large energy gap between the interacting orbitals. Note also that in each case the bonding interaction, e.g. (i), is smaller than the antibonding interaction, e.g. (ii). However, since in each case only two electrons are involved, and these enter the new bonding orbital, the overall effects are stabilising in both cases.

1 cal = 4.184 J

energy levels of the orbitals in each are gradually altered. When the collision leads to chemical change, the orbital energies can, in principle, be tracked as the transition state is approached. With an early transition state, the changes which will have occurred may be viewed as a relatively small perturbation of the orbitals (and their energies) of the reacting partners. The overall effect will almost always be antibonding (hence the activation energy which must be provided for reaction to occur), but it turns out that a very important determinant of activation energy is the interaction between the HOMO of one reaction partner and the LUMO of the other, and that when two related systems are compared, it is this which is the principal factor controlling relative reactivities in the two systems. Of course, for each pair of reaction partners there will be two such interactions, but usually the orbital energies of the partners are such that the HOMO of one is much closer to the LUMO of the other than *vice versa*; since orbital interaction energies are greatest for orbitals of comparable energy, the alternative LUMO–HOMO interaction is relatively unimportant (Fig. **9.1**).

This introduction is hugely oversimplified. It has completely disregarded coulombic interactions which may be of over-riding importance in interactions between charged or highly polarised species. It has also disregarded one essential aspect of frontier orbital theory, namely orbital overlap, the extent of which is also a major determinant of frontier orbital interactions. The interaction with a π-frontier orbital of a conjugated system is likely to be greatest at that site for which the coefficient of the atomic orbital in the MO is greatest. For example, the LUMO of acrolein has atomic-orbital (AO) coefficients indicated in Fig. **9.2**. If the attack of a nucleophile on acrolein is governed predominantly by frontier orbital interaction between this LUMO and the orbital on the nucleophile containing the lone-pair electrons, it will be at the β-carbon atom. As is well known, this is sometimes observed, although many charged nucleophiles attack at carbonyl carbon where the coulombic effect is greatest.

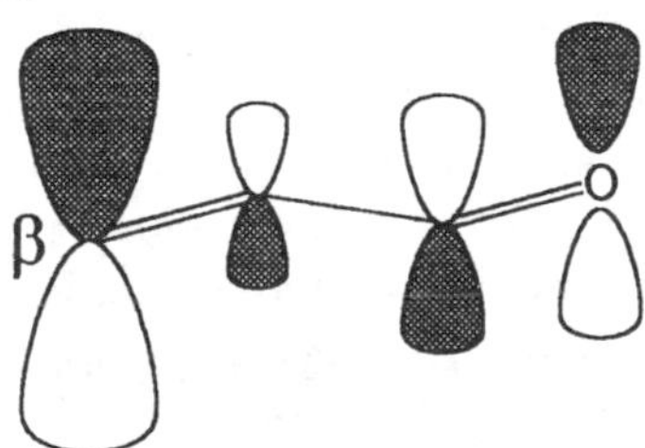

Fig. 9.2: An indication of *p*-orbital contribution to the π-antibonding molecular orbital (LUMO) of acrolein, CH_2=CHCHO.

For radical reactions, the above generalisations require further elaboration. In the first place, because both reaction partners are most usually electrically neutral, coulombic effects are likely to be small. Secondly, for a radical–molecule (i.e. chain-propagating) reaction, the important singly-occupied molecular orbital (SOMO) of the radical may well lie roughly mid-way in energy between the HOMO and LUMO of the substrate. Therefore, both SOMO–HOMO and SOMO–LUMO interactions may have to be considered (Fig. **9.3**). *Both* of these interactions contribute a reduction in energy, i.e. are indicative of a lowering in transition state energy – the former by $(2\varepsilon_1 - \varepsilon_{-1})$, the latter by ε_2. Which of them dominates can be

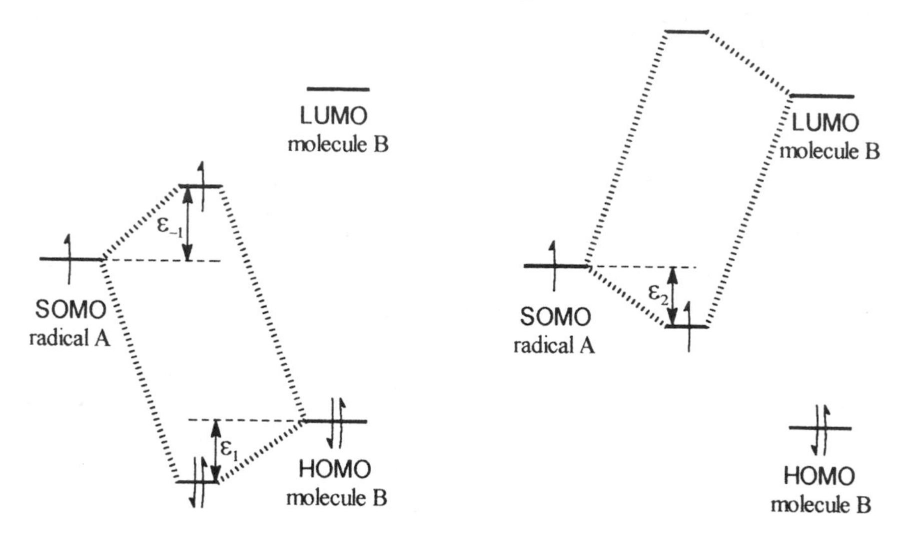

Fig. 9.3: Interaction of the singly occupied orbital of a radical A with the HOMO of molecule B (left-hand diagram), and with the LUMO of molecule B (right-hand). Both interactions are bonding, the former has stabilisation energy $[2\varepsilon_1 - \varepsilon_{-1}]$, the latter ε_2.

related to an alternative description of polar effects (Chapter 4). If the SOMO resides principally upon an electronegative atom, e.g. the oxygen in HO· or RO·, this will lower the SOMO energy so that the SOMO–HOMO interaction becomes dominant; the consequence is that the radical behaves like an electrophile. If, on the other hand, the SOMO energy of the radical is closer to the energy of the LUMO of the substrate, the radical will behave like a nucleophile. Evidently, this type of analysis would be applicable to the two radical addition steps in Scheme 4.3. The attack of the simple alkyl radical on acrylonitrile is facilitated by the reduction in LUMO energy of the substrate by the cyano group, so that the SOMO–LUMO interaction dominates; furthermore, because the highest AO coefficient of the acrylonitrile LUMO is, like that of acrolein, on the β-carbon atom, addition occurs at that site. In contrast, the electron-withdrawing effect of the cyano-group in the resulting radical is to lower the SOMO energy. The HOMO energy of an alkene is known to be raised significantly by aryl-substitution, so that the principal interaction between the cyanoalkyl radical and the α-methylstyrene is now of the SOMO–HOMO kind – the new radical is behaving as an electrophile. There are many examples of similar behaviour to be found in radical copolymerisation systems,

1 cal = 4.184 J

where mixtures of vinyl monomers, one of which has an electron-withdrawing substituent, the other an electron-donating one, will form polymers incorporating both monomers in an alternating sequence (**1**).

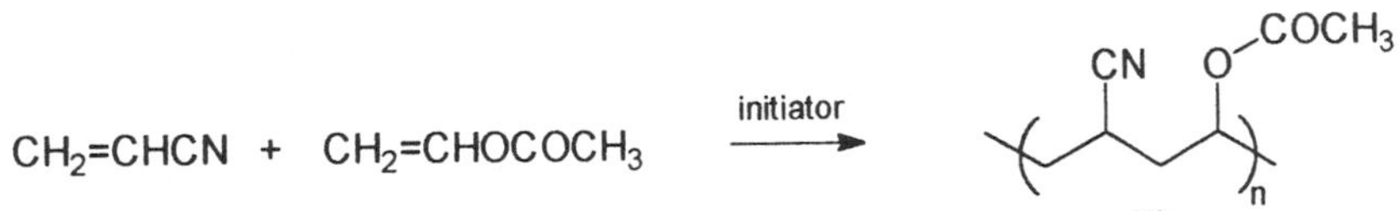

$$CH_2{=}CHCN + CH_2{=}CHOCOCH_3 \xrightarrow{\text{initiator}}$$

CN O COCH₃ n

(**1**)

9.2 RADICAL KINETICS AND THERMODYNAMICS

Devising methods for determining absolute rate data or thermodynamic properties for species which cannot be isolated, and which may be difficult to accumulate in concentrations greater than, say, 10^{-8} molar has inevitably afforded many difficulties, and methods for their solution are still emerging. Kinetic ESR techniques have been mentioned elsewhere, but the difficulties associated with accurate measurement of radical concentrations have hampered direct measurement of bimolecular rate constants. Nevertheless, a careful study of the bimolecular decay of t-butyl (by disproportionation and dimerisation) in heptane provides an example which has given a second-order rate constant ($2k_t$ ca. 8×10^9 M^{-1} sec^{-1} at 20°C) similar to the diffusion-controlled limit in that solvent. Although techniques of flash photolysis and pulse radiolysis can generate relatively high concentrations of transients, and the associated kinetic spectroscopy can readily embrace time-scales appropriate for studying the decay of those transients, many small radicals, for example, simple alkyl radicals, are optically transparent at available monitoring wavelengths; they must therefore be studied by monitoring product radicals or by other more indirect means. This is illustrated by experiments on the reactions of perfluoroalkyl radicals. The radicals were generated by laser flash photolysis of perfluoroalkanoyl peroxides [$(R_fCO_2)_2$] in the presence of known concentrations of, for example, β-methylstyrene in an unreactive solvent (Freon 113). From the pseudo-first-order growth of the absorption due to the benzylic radical adduct (**2**) the rate of addition to the alkene could be calculated. The experiments were extended to obtain rate data for addition to non-conjugated alkenes such as 1-hexene. But in this case the radical adduct, being a simple alkyl radical, is optically transparent, so that rates were estimated by

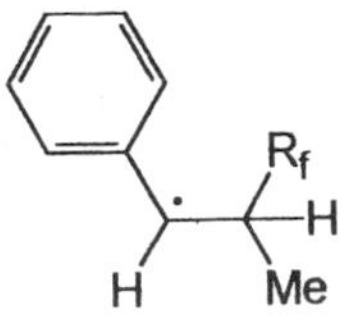

R_f H H Me

(**2**)

using β-methylstyrene as a spectroscopic "probe" in which it was allowed to compete with the hexene for fluoroalkyl radicals. The rate of addition to hexene is then deduced from the reduction in the amount of benzylic radical adduct which is formed.[1]

Application of very short time scale flash-photolysis techniques has in some instances permitted direct observation of radical cage phenomena.

As with kinetic parameters, measurements of thermodynamic parameters have been difficult, and, as mentioned in Chapter 4, bond dissociation energies and related heats of formation of radicals have been the subject of frequent revision in the chemical literature. Early measurements on the thermodynamics of simple alkyl radicals depended on gas-phase studies of the kinetics of alkane iodination, but quite serious systematic errors have been identified in these, so that C–H bond dissociation energies and heats of formation were consistently too low. The data have been reexamined, and revised figures are now in reasonable agreement with results from other techniques. Interestingly, data obtained by modern high-level MO calculations may compare in accuracy with the best experimental results.

A simple approach to estimating the resonance stabilisation of a number of conjugated radicals has depended on ESR observations of the rates of conformational change when this requires rotation through a conformation in which conjugation is broken. An example of this is given below under the heading of captodative stabilisation.

The strengths of some weaker X–H bonds, e.g. S–H, have been estimated by photo-acoustic calorimetry. These experiments measure the heat of reaction of RS–H with t-butoxyl which is generated by flash photolysis of di-t-butyl peroxide. The S–H bond strength estimate is in effect based on an estimate of the strength of O–H in t-butanol.[2] This indirect approach to thermochemical data mirrors the indirect approach to so much of the accumulated kinetic data which are now available (see Appendix).

Consideration of the approach to reactivity in alkane halogenation which was presented in Chapter 4, may have left the reader wondering whether there might be a simple quantitative relationship between the rates of the atom-transfer steps and their heats of reaction. The first, and best-known relationship of this kind is the Evans–Polanyi equation (1).

$$E_a = E_o + \alpha\Delta H \qquad (1)$$

[1]In some instances, preparative experiments showed that allylic abstraction competes with addition. It these cases a small correction factor must of course be applied when calculating rate constants for addition steps.

[2]This can be deduced from a simple thermochemical cycle depending on the known O–O bond dissociation energy of the peroxide (Chapter 3), together with the heats of formation of t-butanol, di-t-butyl peroxide and H·.

1 cal = 4.184 J

This correlates the activation energy E_a with the heat of reaction ΔH; E_o and α are constants. Whilst the Evans–Polanyi relationship is reasonably satisfactory for a series of closely related reactions, it fails totally to correlate, for example, hydrogen atom transfers of disparate types. Brief consideration of polar effects suggest that such a breakdown is to be expected. Although the most recent molecular orbital calculations on reaction trajectories reproduce activation energies of radical reaction steps with some precision, one recent attempt to derive a purely empirical relationship applicable to a wide range of hydrogen-atom transfers of the form A· + H-B $\rightarrow$ A-H + B· does give excellent agreement with data for some sixty individual reactions (correlation coefficient > 0.98). Furthermore, the relationship [Equation (2)] is interesting in that the additional terms can be related pictorially to variations between individual reaction systems.

$$E_a = E_o f + \alpha \Delta H(1 - d) + \beta(\Delta\chi)^2 + \gamma(s_A + s_B) \qquad (2)$$

In this equation, in which β and γ are new constants, $\Delta\chi$ is the difference between the Pauling electronegativities of A and B, the third term therefore being introduced to allow directly for polar effects. The s terms are introduced to allow for differences in geometry as A and B move from reactants to transition state. For example, if A is a planar methyl radical, it will be pyramidalised in the transition state. In a somewhat similar manner, d values allow for changes in conjugative stabilisation on proceeding from some reactants, e.g. toluene, to transition state. Finally, f values take account of the absolute bond strengths of of the bonds being broken and made; this allows for the intuitive expectation that two thermoneutral reactions would not have the same activation barrier if the strengths of the bonds undergoing change in the two processes are very different – in contrast to any prediction based on Equation (1).

9.3 CAPTODATIVE STABILISATION

A C–H bond is usually weakened by an adjacent electron-donating or -withdrawing group such as alkoxyl or carbonyl, but the effect is generally quite small, perhaps 3–5 kcal mol^{-1}. It has been suggested that when both types of substituent are present there is a synergistic effect between them, and that radical stabilisation is greater than might be expected based on the sum of the effects of the two separate substituents. The effect has been referred to as "captodative" stabilisation, where the electron-withdrawing group is the "capto" group, and the electron-donating group is the "dative" one, e.g. **(3)**. The idea has prompted extensive experimentation, and there is evidence to support the idea in that there does seem

MeO–Ċ(H)–CN

(3)

to be a kinetic (i.e. *transition state*) effect of this kind, but there are few experiments which provide evidence for any special *thermodynamic* stabilisation of radicals by this effect. One, which apparently does, illustrates the ESR approach mentioned in the previous section. Analysis of broadening of spectra of α-aminoalkyl radicals **(4)** due to rotation of the amino group indicates a barrier of 7.5 kcal mol^{-1} when R = Me, which rises to 11 kcal mol^{-1} when R = CN. This is consistent with extended conjugation in the latter case.

$H_2N\dot{C}(R)H$

(4)

In the more extreme (as yet experimentally unknown) example, (**4**; R = BH_2), high level *ab initio* molecular orbital calculations have lent some theoretical support to the captodative effect. The stabilisation energies of $\cdot CH_2BH_2$ and $\cdot CH_2NH_2$ (relative to $\cdot CH_3$) were both calculated to be ca. 10 kcal mol^{-1}, whereas that of (**4**; R = BH_2) was 33 kcal mol^{-1}. The estimated captodative stabilisation of 13 kcal mol^{-1} is (fortuitously?) remarkably close to the resonance energy of the isoelectronic allyl radical (usually quoted as ca. 14 kcal mol^{-1}).

9.4 AUTOXIDATION OF UNSATURATED LIPIDS

Although the basic principles of autoxidation have been encountered earlier (Chapters 2 and 4), there is one context which merits special consideration because of both its exceptional significance and its relative complexity. This concerns the autoxidation of unsaturated fatty acids that are incorporated into natural oils and fats, and are therefore present in foodstuffs, and in living plant and animal tissues. Their oxidation by molecular oxygen is responsible for the production of "off-flavours" and rancidity in foods that have been subjected to poor or prolonged storage conditions. It is responsible for the hardening of "drying oils", such as linseed oil, that were the basis of many traditional paint formulations, and, as we shall see in the final chapter, it has considerable significance in living organisms.

Long-chain fatty acids $R_{1\text{-}c}CO_2H$ are abundant in nature, most commonly in combination with glycerol either as triglycerides (**5**) or as the amphiphilic membrane phospholipids such as phosphatidylcholine (lecithin) (**6**). From most sources, either (**5**) or (**6**) will be found to comprise a mixture of compounds with differing proportions of saturated ($C_nH_{2n+1}CO_2$-), mono-unsaturated ($C_nH_{2n-1}CO_2$-), and poly-unsaturated fatty acid (PUFA) residues. In the case of (**5**), a high degree of saturation is likely to render the mixture solid at room temperature – a fat; high proportions of mono- or poly-unsaturated residues produce liquids under the same conditions – the oils. Particularly common are fatty acid residues with eighteen carbon atoms (Fig. **9.4**), and we shall especially focus our attention on derivatives

1 cal = 4.184 J

of linoleic and linolenic acids.[1] Both of these incorporate "skipped diene" units (–CH=CH–CH_2–CH=CH–). Reference to Table 4.3 will remind the reader that allylic C–H bonds are relatively weak; the C–H bonds of the central methylene group in skipped dienes are "doubly allylic" and are especially weak (ca. 75 kcal

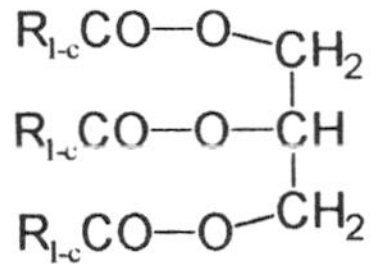

(5)

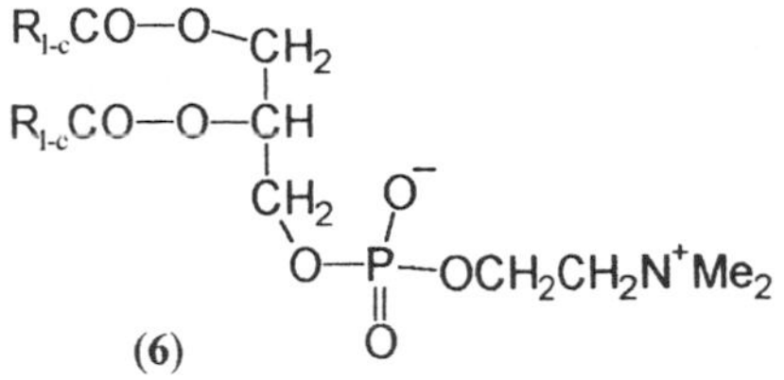

(6)

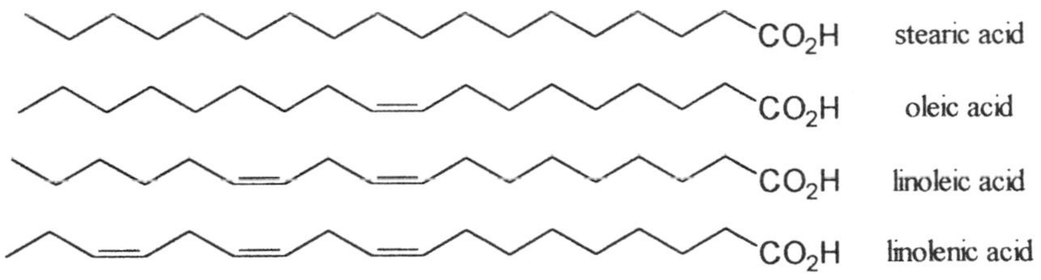

stearic acid

oleic acid

linoleic acid

linolenic acid

Fig. 9.4: Some C_{18} long-chain fatty acids $R_{1\text{-}c}CO_2H$ which are commonly found as glyceryl esters. Note the *cis*-configuration of the double bonds.

mol^{-1}). They are consequently unusually susceptible to radical attack, removal of a hydrogen atom giving resonance-stabilised dienyl radicals [Equation (3)]. Any peroxyl radical, which is a relatively unreactive species, will selectively abstract hydrogen from these doubly allylic sites.

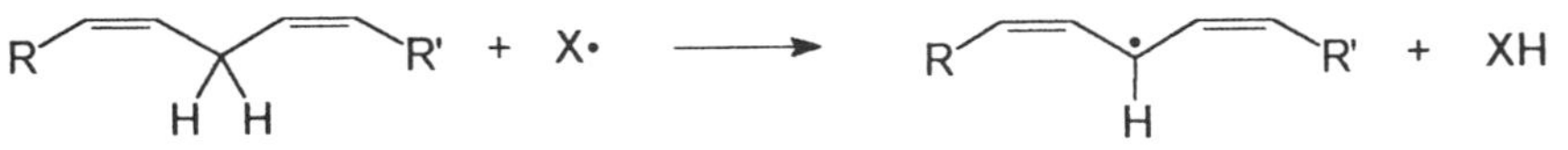

(3)

Whilst it is therefore plausible that radical-chain autoxidation of structures incorporating skipped dienes should be very facile, one further consideration is important, namely that the dienyl radicals derived from them will react with oxygen almost exclusively at a terminal carbon to yield a peroxyl radical in which the two double bonds have moved into conjugation. These principles are summarised in Scheme 9.1, which also notes stereochemical features of the process.

When a pre-weighed thin film of linseed oil on glass is exposed to the air for several hours there is a perceptible weight increase. Elemental analysis confirms that this is due to the incorporation of oxygen. Further weight increase is

[1]Note that in the majority of fatty acid residues the double bonds have the *cis* configuration.

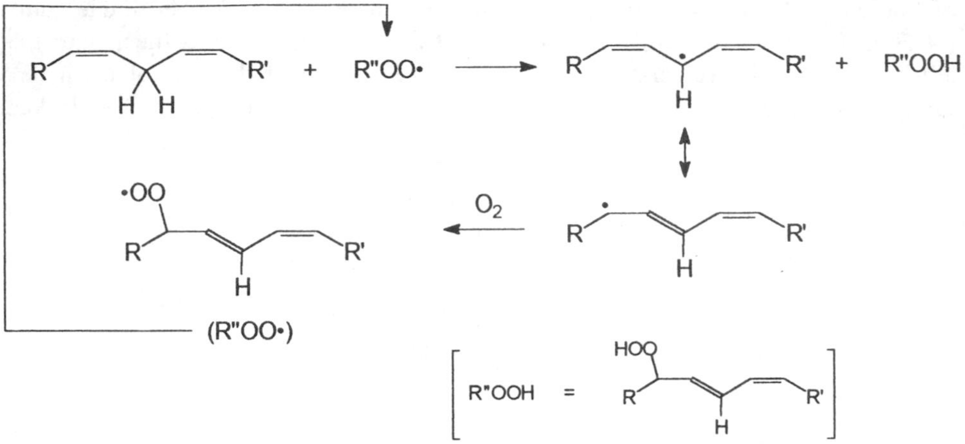

Scheme 9.1: Chain-propagating sequence for autoxidation of a skipped diene. Note that in the derived hydroperoxide (R"OOH) the newly positioned double bond is predominantly *trans*.

accompanied by an increase in viscosity of the oil film, which eventually starts to harden. This is the "drying" of a drying oil, a process which does *not* involve solvent loss. At this stage, continuing oxidation generates peroxyl radicals, some of which react by addition to the conjugated diene units that have been formed in the initial oxidation. In this way the oil molecules are linked together and eventually become highly polymerised. However, after very prolonged exposure to air, the rate of gain in weight not only falls, but is eventually replaced by weight loss. This can come about by peroxide or hydroperoxide decomposition to generate alkoxyl radicals which may fragment, releasing small molecules, typically aldehydes, which contribute to the odour of drying paint or deteriorating foods. A plausible example is given in Scheme 9.2, but detailed analysis of volatiles liberated from triglyceride autoxidation reveals an astonishingly complex pattern of behaviour. This chemistry is further complicated by the presence of transition metal ions. These can obviously catalyse hydroperoxide decomposition (see Scheme 9.2), but a complete rationalisation of observed effects of different metals is lacking. The importance of metal ions is reflected in the addition of lipid-soluble transition metal carboxylates ("driers"), particularly those of cobalt and manganese, to modern paint formulations, many of which still depend on autoxidation processes for setting and hardening. However, the history of these effects may be traced back to the easel paintings of the Old Masters. Jan van Eyck (d. 1441) is commonly credited with being the first to mix what we now know to have been lead compounds with other mineral pigments in order to accelerate the drying of his paints, and it has long been known that oils containing those blue or green pigments which we now recognise as incorporating copper salts were particularly slow to harden.

1 cal = 4.184 J

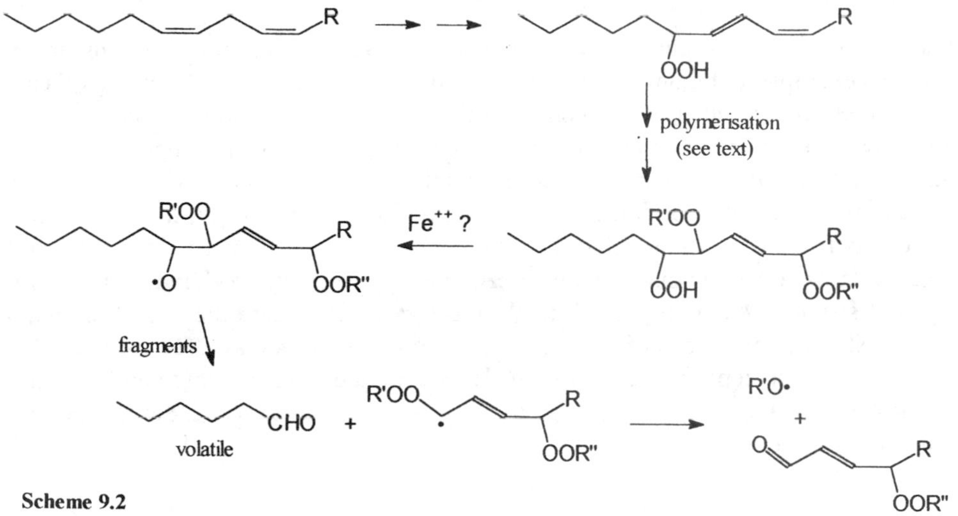

Scheme 9.2

There are other complications, too. For instance, it has been demonstrated that dienylperoxyl radicals, e.g. **(7)**, are formed reversibly. This affords, *inter alia*, a mechanism for double-bond isomerisation during autoxidation (Scheme 9.3). During the life-time of the dienyl radicals themselves however, there is sufficient double-bond character between each pair of adjacent carbon atoms that the radicals tend to be conformationally stable.

We shall return to lipid autoxidation in the final chapter.

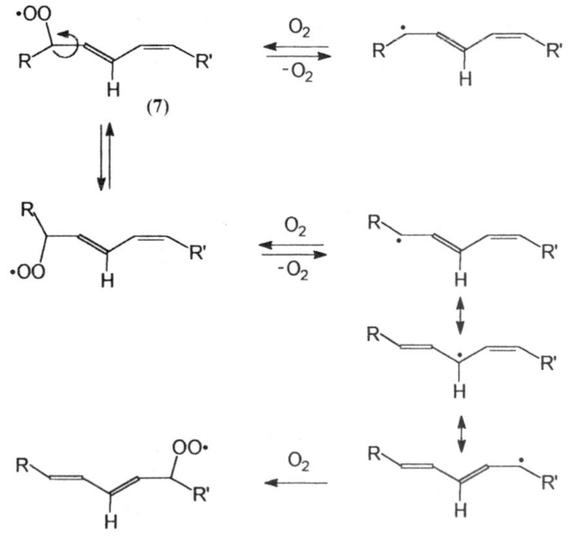

Scheme 9.3: Genesis of an all-*trans* diene by reversibility of dienylperoxyl formation.

9.5 ANTIOXIDANTS

Oxidation of many organic substances can be a serious nuisance. Foods are an obvious example, but many commercial products such as rubber, lubricating oils and diesel fuel are degraded by oxygen from the air. These compounds may not have such reactive centres as the labile methylene hydrogens of a skipped diene, but allylic or benzylic hydrogens and tertiary aliphatic hydrogens are among groupings which are susceptible to radical chain autoxidation.[1] Fortunately, there are means of combating this, chief amongst which is the addition of "antioxidants". Antioxidants are compounds which react rapidly with peroxyl radicals to form product species which are insufficiently reactive to propagate the oxidation chain. The most commonly used of these are hindered phenols such as BHT (**8**) and BHA (**9**).[2] These intercept the peroxyl radicals to generate hydroperoxide and hindered phenoxyl radicals, and the latter are then capable of scavenging a second peroxyl radical (Scheme 9.4).

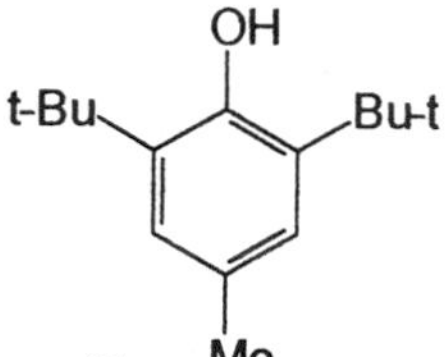

OH
t-Bu Bu-t
Me
(**8**)

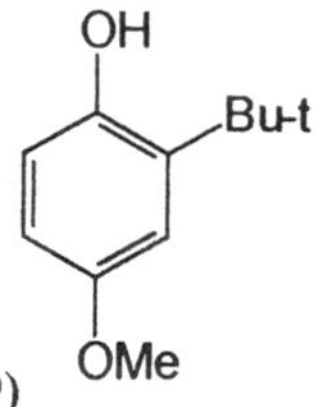

OH
Bu-t
OMe
(**9**)

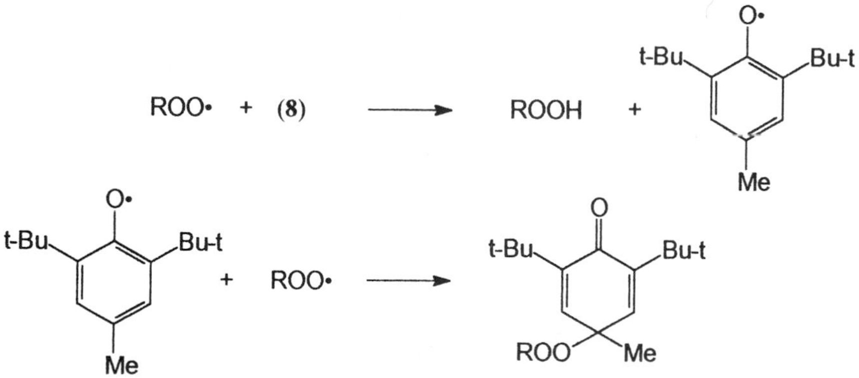

ROO• + (**8**) ⟶ ROOH + [O•, t-Bu, Bu-t, Me]

[O•, t-Bu, Bu-t, Me] + ROO• ⟶ [O, t-Bu, Bu-t, ROO, Me]

Scheme 9.4

[1]Oxidative degradation of unsaturated hydrocarbons by singlet oxygen is also a serious problem.
[2]Compound (7), 2,6-di-t-butyl-*p*-cresol, is commonly referred to as BHT (butylated hydroxytoluene), and (**8**), 2-t-butyl-4-methoxyphenol, is known as BHA (butylated hydroxyanisole).

1 cal = 4.184 J

9.6 ADAMANTANE MATRIX STUDIES

It was mentioned in Chapter 5 that reactive radicals could be stabilised in solid matrices at low temperatures for study by ESR, and this was exemplified in Chapter 8. It must be remembered, however, that radicals in this type of matrix are fixed in random orientations with respect to the magnetic field, so that anisotropic broadening usually dominates the appearance of spectra which have been obtained in this way, often making interpretation difficult. Spectroscopists have explored various ways of immobilising radicals with respect to bimolecular collision whilst at the same time retaining freedom of rotational motion and intramolecular reorganisation. Several approaches have met with varying degrees of success, among them inclusion in zeolites, and incorporation into micellar systems, where quite long life-times have been observed for radicals generated at concentrations lower than the equivalent of one radical per micelle. Phenoxyl radicals from antioxidants such as BHT have been readily detected by ESR in autoxidised *cis*-polybutadiene, where clearly resolved isotropic spectra indicate almost complete rotational freedom, but where diffusion through the polymer matrix must, at the very least, be slow; other polymer systems have afforded similar results. However, one of the most productive strategies has been to use an adamantane matrix. This incorporates cavities between the adamantane molecules which are quite large enough to include molecules of modest dimensions, e.g. toluene, but to block them from diffusing through the lattice. Radiolysis of such entrapped molecules has permitted some remarkable observations to be made. As well as yielding good isotropic spectra of, for example, benzyl radicals, radical rearrangements which cannot be followed in solution may be directly monitored. For example, ESR studies of hydrogen abstraction from bicyclo[3.1.0]hex-2-ene in liquid solution have revealed only the bicyclohexenyl radical (**10**). X-Irradiation of the bicyclohexene in an adamantane matrix at room temperature gives isotropic spectra of cyclohexadienyl. At much lower temperatures, only (**10**) is seen. However at ca. –50°C the interconversion may be directly monitored, with an estimated activation barrier ($\Delta G^{\ddagger}$) of ca. 14.5 kcal mol^{-1}. This is too high for the reaction to occur to any significant extent during the life-time of (**10**) in liquid solution. Deuterium labelling further demonstrated that a faster sigmatropic isomerisation of (**10**) also occurs [(**10**) ⇄ (**10'**) etc.].

[- H·]
H
(**10**) H
(**10'**)
H
H

Despite the evident significance of this procedure, it is important to recognise that rate data obtained from any matrix studies should be extrapolated to the more familiar environment of liquid solution with extreme caution.

9.7 THERMOLYSIS OF DIBENZOYL PEROXIDE IN BENZENE

We saw in Chapter 1 how results of aromatic substitution reactions led to the discovery that reactive radicals can mediate in some solution processes. These substitution reactions involved aryl radicals, and several sources of these were investigated in those early experiments. With any monosubstituted benzene, the substitution patterns observed were independent of the source of the aryl radical, yet the yields of biphenyl derivatives were often poor, and it was clear that detailed understanding of the reactions was lacking. One much-studied precursor of phenyl radicals was benzoyl peroxide (dibenzoyl peroxide; **11**). This was extensively employed as an initiator long before the complexities of its reaction with benzene were better understood – in the early 1960s. The story is an interesting one, and some of the more significant results and ideas which led to our current understanding of the problem are outlined below. Unless indicated otherwise, all experiments were carried out in the absence of oxygen.

$$PhC(=O)OOC(=O)Ph \ (\mathbf{11}) + C_6H_6 \longrightarrow PhCO_2H + PhC_6H_5 + CO_2 \quad (4)$$

The early discussion of the thermolysis of dibenzoyl peroxide in benzene was represented in terms of Equation (4), and was deduced to proceed *via* benzoyloxyl and phenyl radicals [(**11**) → $PhCO_2\cdot$ → $Ph\cdot + CO_2$]. However, it was soon recognised that this stoichiometry provided a very poor description of all that must be going on, since the yields of biphenyl and benzoic acid were seldom more than ca. 40%, whereas the yield of carbon dioxide was more like 150%! In addition, a large quantity of unidentified tarry product was obtained. A further curiosity was the fact that a small quantity of the rather insoluble *p*-quaterphenyl (**12**) separated from the reaction mixtures. Although this was not a particularly significant product, it seemed to be formed in far too great a yield to be accounted for in terms of the sequence outlined in Equation (5), unless the intermediates biphenyl and *p*-terphenyl were *very* much more reactive towards Ph· than was benzene.

$$PhC_6H_5 \xrightarrow{Ph\bullet} PhC_6H_4Ph \xrightarrow{Ph\bullet} \underset{(\mathbf{12})}{Ph(C_6H_4)_2Ph} \quad (5)$$

1 cal = 4.184 J

The solution to this benzoyl peroxide problem was unduly long in coming, perhaps because the early investigators were preconditioned by their knowledge of electrophilic substitution. The difference, which they failed at first to recognise, is that the Wheland intermediate of, say, nitration is rapidly deprotonated by weak bases present in the reaction medium, whilst the equivalent intermediate in radical substitution is a delocalised cyclohexadienyl radical **(13)** in which the sp^3-bonded hydrogen is held too strongly for spontaneous dissociation to occur. Instead, the radical behaves like any other carbon-centred radical and reacts *inter alia* by dimerisation and disproportionation. The products of such reactions would be rather labile hydroaromatic species, such as **(14)**, **(15)** and **(16)**. And in the case of the dimer **(15)**, this is further complicated by the possibility of positional and

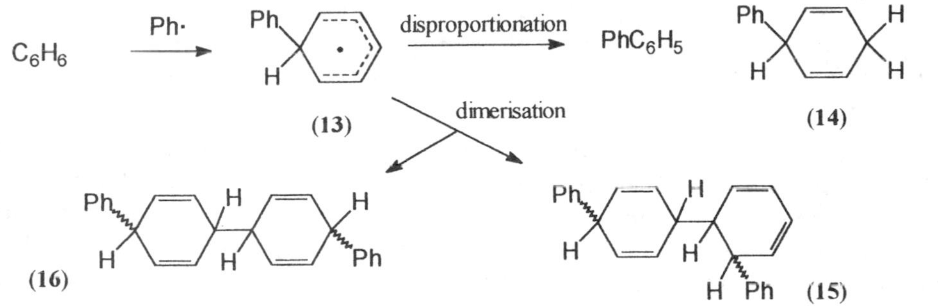

C_6H_6 —Ph·→ (13) —disproportionation→ PhC_6H_5 + (14)

(13) —dimerisation→ (16) + (15)

stereoisomerism. The proposed products all contain doubly allylic hydrogens and may incorporate conjugated double bonds (e.g. **15**), both of which are very much more susceptible to radical attack than is the benzene solvent. To minimise the complications which might therefore be expected to arise as a result of secondary reactions, the thermolysis was carried out in very dilute solution. From such experiments both 1,4-dihydrobiphenyl **(14)** and one stereoisomer of the dimer **(16)** were isolated. In order to eliminate the stereoisomer problem, the products of one of these high dilution experiments were subjected to dehydrogenation using *o*-chloranil **(17)**; from this experiment all three expected quaterphenyls **(12)**, **(18)**, and **(19)** were isolated. Under these high dilution conditions, it was also noted that the yield of benzoic acid was much reduced.

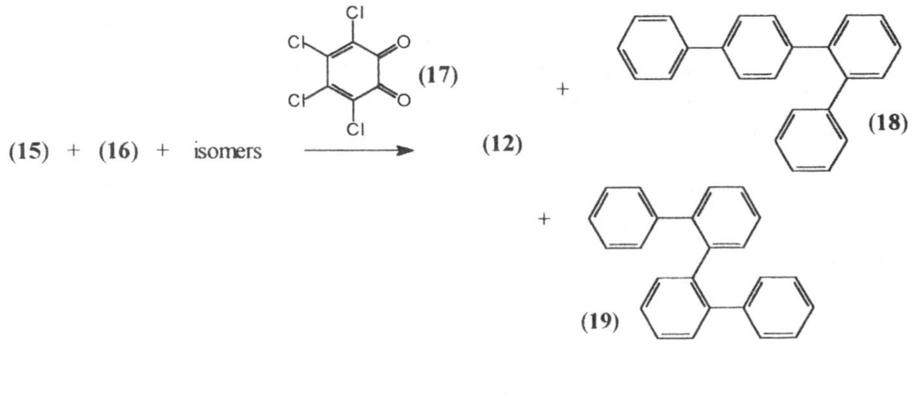

(15) + **(16)** + isomers —(17)→ **(12)** + **(18)** + **(19)**

The reasonable assumption might then be that when higher concentrations of peroxide are decomposed the reactive hydroaromatic compounds are partially converted to an even more complex mixture, but that because transient radical concentrations will be higher more benzoyloxyl radicals will react with **(13)** prior to decarboxylation. This would account for the increased yield of benzoic acid. However, in parallel with the product studies, kinetic investigations were carried out from which it was demonstrated that as the initial concentration of benzoyl peroxide, $[P]_o$, was increased, so the fraction which decomposed by a radical-induced mechanism increased also. The kinetic evidence pointed very strongly to the species responsible for the induced peroxide decomposition being the intermediate **(13)**, as indicated in Equation (6). Furthermore, a close parallel was found between the yield

$$\textbf{(13)} + PhC(=O)OOC(=O)Ph \longrightarrow PhCO_2H + PhC_6H_5 + PhCO_2\bullet \qquad (6)$$

of benzoic acid as a function of $[P]_o$ and the extent of induced decomposition deduced from the kinetics (Fig. **9.5**). This strongly suggested that the principal source of benzoic acid is in fact the induced decomposition. The small discrepancy between the curves in the figure is best accommodated by assuming that one of the secondary reactions involves abstraction of the doubly allylic hydrogens by benzoyloxyl radicals (thus explaining the small quantities of quaterphenyl seen in the early investigations at relatively high $[P]_o$).

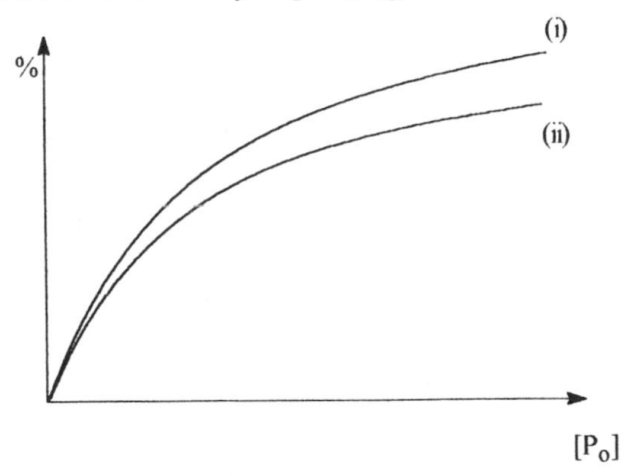

Fig. 9.5: Graphs indicating (i) the percentage yield of benzoic acid according to Equation (4), and (ii) the percentage of peroxide decomposing by a radical-induced pathway, both plotted as a function of initial benzoyl peroxide concentration.

Not all of the puzzles associated with this chemistry have been resolved. For example, decomposition of the peroxide in chlorobenzene gives chlorobiphenyls in no more than 50% yield, the explanation being essentially the same as for the

1 cal = 4.184 J

benzene case. However, the biaryl yields may be greatly increased by bubbling oxygen through the system. The cyclohexadienyl radicals are then efficiently oxidised, and little or no tetracyclic (dimeric) product is formed. The surprising result, in the case of chlorobenzene, is that this modification is without influence on the proportions of the three chlorobiphenyl isomers. In other words, contrary to what might have been expected on the basis of, for example, steric factors, in the oxygen-free system the partitioning of each of the three isomeric chloro-cyclohexadienyl intermediates between biphenyl formation on the one hand and dimerisation on the other must be almost identical.

9.8 STABLE-RADICAL EFFECTS

From time to time radical reactions have been discovered in which products isolated are formed by cross-coupling of two different radicals *unaccompanied by products of symmetrical coupling*. This was sometimes presented simply as a "preference for unsymmetrical coupling". In other cases it led to the hypothesis that some extreme form of cage behaviour must be operative. We have already encountered some examples. In the Toray process (Chapter 6), bicyclohexyl is a negligible product, and from the Barton aldosterone synthesis (Chapter 6), dimeric steroids have not been isolated.

In another example, the phenyl radical precursor, phenylazotriphenylmethane (**20**), when allowed to decompose thermally in boiling carbon tetrachloride, gives near quantitative yields of chlorobenzene and 1,1,1-trichloro-2,2,2-triphenylethane. Hexachloroethane, the frequently encountered dimer of $CCl_3\cdot$, is apparently not formed. In benzene this azo-compound gives biphenyl, triphenylmethane, and the cross-coupling products (**21**); neither dihydrobiphenyls nor the cyclohexadienyl dimers discussed in the previous section could be detected.

$$\underset{\textbf{(20)}}{PhN{=}NCPh_3} \xrightarrow{CCl_4} N_2 + PhCl + Cl_3CCPh_3$$

$$PhN{=}NCPh_3 \xrightarrow{C_6H_6} N_2 + PhC_6H_5 + HCPh_3$$

+ (Ph)(H)C-cyclohexa-2,5-diene-C(H)(CPh3)

(**21**) *cis* and *trans*

The common feature in all these reactions is the participation of a stable radical (NO or triphenylmethyl). Taking the example of phenylazotriphenylmethane in CCl_4, the azo-compound decomposes to form phenyl and triphenylmethyl radicals at an equal rate. The reactive phenyl radicals remove chlorine from the solvent to produce the rather less reactive $CCl_3\cdot$. Should some of these dimerise, this would necessarily result in the build up of a reservoir of triphenylmethyl radicals (in equilibrium with the dimer; see Chapter 1). Subsequently, since the reactions of $CCl_3\cdot$ both with itself and with triphenylmethyl will be essentially diffusion-controlled, the one of these two reactions that will successfully lead to product will be that involving the far more abundant triphenylmethyl radical. As the reaction proceeds, this does *not* deplete the reservoir of triphenylmethyl, because for every $CCl_3\cdot$ radical which is produced, another triphenylmethyl is added to the reservoir. The concentration of triphenylmethyl during these experiments is sufficient to give a strong ESR signal. The trichloromethyl radical is not detectable. The process is outlined in Scheme 9.5

Although this stable-radical effect was correctly rationalised by several authors in the 1960s, it is only relatively recently that it has been subjected to detailed kinetic analysis.

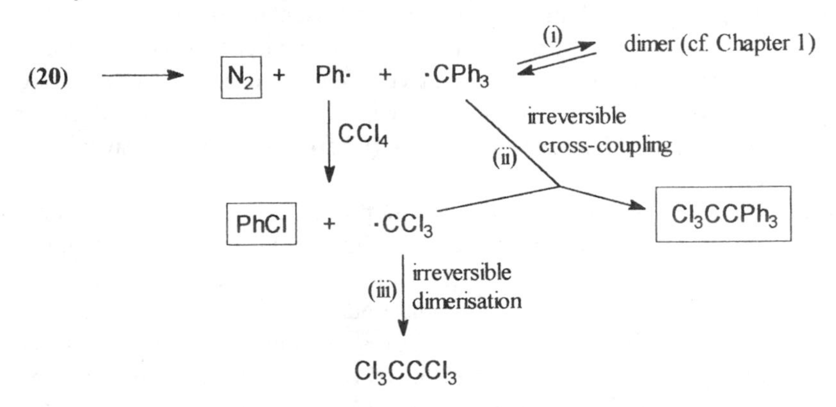

Scheme 9.5: Steps (i) and (ii) are irreversible and have similar (diffusion-controlled) rate constants. Therefore trace occurrence of (iii) will result in a much higher stationary state concentration of $\cdot CPh_3$ than $\cdot CCl_3$, since the former dimerises reversibly. Subsequently, $\cdot CCl_3$ will react almost exclusively with this high concentration of $\cdot CPh_3$. The observed products are those enclosed in boxes.

9.9 RADICALS FROM KETONE PHOTOCHEMISTRY

Space limitations have in general precluded any major excursions into organic photochemistry, but two aspects of the excited-state reactions of ketones encroach upon our territory. In the first place, ketone photolysis has provided a classical source of some radicals by homolysis of the bond α to the carbonyl group [Equation (7)]. Many early studies of radical behaviour were carried out in the gas phase, where, for example, photolysis of acetone yields methyl radicals. The quantum efficiency of that reaction is tiny in solution, but carbonyl compounds in which the

1 cal = 4.184 J

carbonyl–alkyl bond is weaker such as di-t-butyl ketone, are readily photolysed in the liquid phase.

$$\mathrm{RC(O)R} \xrightarrow{h\nu} \mathrm{R\cdot} + \mathrm{R\dot{C}{=}O} \quad [\longrightarrow \mathrm{R\cdot} + \mathrm{CO}] \qquad (7)$$

A second type of photoreaction, which is perhaps more directly relevant to a discussion of radical behaviour, involves aryl ketones, typically benzophenone,[1] in which absorption of ultraviolet light is followed by very rapid electron-spin inversion to generate a triplet state species. The structure of this can be considered as arising by the transfer of one electron from an oxygen lone-pair ("n" or non-bonding) orbital into a π^*-antibonding orbital which embraces the whole π-system of the molecule. This species, referred to as an n,π^* triplet excited state, has radical reactivity very similar to that of an alkoxyl radical. The result is unsurprising, since both have a single electron localised in a non-bonding orbital on the oxygen atom. The reactivity pattern is particularly reflected in hydrogen abstraction reactions, e.g. Equation (8). The derived radicals behave similarly to radicals generated by alternative procedures, except that, since electron spin is conserved in the atom transfer step, the initially formed radical pair will, like its ketone precursor, be in the triplet state, i.e. with the electron spins of the single electrons parallel. This clearly has consequences for CIDNP effects (Chapter 5), since spin inversion must precede radical coupling or disproportionation.

$$\mathrm{PhC(O)Ph} \xrightarrow{h\nu} \longrightarrow \mathrm{Ph_2\overset{\uparrow}{C}{-}\overset{\uparrow}{O}} \xrightarrow{\mathrm{RH}} \mathrm{R\cdot} + \mathrm{Ph_2\dot{C}{-}OH} \qquad (8)$$

Unsurprisingly, many reactions of ketone triplets have been investigated in which *intra*molecular hydrogen transfer occurs. The immediate result is the formation of a biradical which can collapse in a variety of competing processes. Study of these, using techniques which range from product analysis to picosecond laser photolysis and kinetic spectroscopy has led to interesting photochemical syntheses, and to fundamental information on the electronic character of ketone excited states. For further information the interested reader is directed to any text on organic photochemistry.

9.10 TRANSITION-METAL IONS IN RADICAL REACTIONS

This topic has been encountered extensively in other sections both of this and other chapters. Nevertheless, a few paragraphs are merited with which to attempt briefly to draw together some of the threads, and to point out some of the complexities.

[1] In which the bonding to phenyl is too strong for α-fission to occur.

We have particularly highlighted Fenton chemistry, but in the example of Scheme 4.3, which requires regeneration of the iron(II) from iron(III), the iron(III) is effectively behaving as a "stable radical" in the context discussed above. Thus there is no coupling of the relatively stable tertiary benzylic radical; instead it is intercepted by the stationary concentration of iron(III).[1] A similar pattern is observed when the thermolysis of benzoyl peroxide in benzene is modified by the presence of a low concentration of cupric benzoate. The peroxide decomposition is accelerated by electron transfer from copper(I), and the resulting copper(II) behaves as a stable radical and intercepts the cyclohexadienyl intermediate **(13)** (Scheme 9.6). This results in near quantitative yields of biphenyl and benzoic acid according to the stoichiometry of Equation (4).

$$PhC(=O)OOC(=O)Ph + Cu(I) \longrightarrow PhCO_2\bullet + PhCO_2Cu(II)$$

$$PhCO_2\bullet \longrightarrow Ph\bullet + CO_2$$

$$Ph\bullet + C_6H_6 \longrightarrow \textbf{(13)}$$

(13)

$$\textbf{(13)} + PhCO_2Cu(II) \longrightarrow PhCO_2H + PhC_6H_5 + Cu(I)$$

Scheme 9.6

Whilst the last example is most simply interpreted in terms of electron transfer, in some reactions of alkyl radicals with copper salts there is evidence for intermediate alkylcopper species formed from alkyl radical and a copper(II) salt, and these can apparently react in different ways according to the nature of the other ligands attached to copper, and to the polarity of the medium. Thus in an ionising medium with very electronegative ligands, such an alkylcopper species may readily undergo solvolysis, effectively *via* R^+. In contrast, with carboxylate ligands, and in a less polar environment, oxidative elimination to alkene usually dominates. The involvement of copper in the oxidative elimination reactions shows up in a marked preference for formation of the least substituted alkene. This is illustrated by the

[1]It is sufficient that $k_2[R\cdot][Fe^{3+}] >> k_t[R\cdot]^2$; the reaction between R· and Fe^{3+} need not be diffusion-controlled.

1 cal = 4.184 J

product of the unusual tandem 6-*endo* cyclisation of farnesyl acetate by benzoyl peroxide in the presence of a copper carboxylate [Equation (9)].

$$PhCO_2\cdot + \text{[farnesyl acetate]} \longrightarrow \longrightarrow \text{[PhCO}_2\text{-bicyclic radical, OCOCH}_3\text{]} \xrightarrow[Cu^{++}]{-H^+} \text{[PhCO}_2\text{-bicyclic alkene, OCOCH}_3\text{]} \qquad (9)$$

When certain main group halides, e.g. lithium chloride or bromide, are added to carboxylic acid oxidations using lead tetraacetate, alkyl halides are formed, often in high yield. No cationic rearrangement of the alkyl group is observed, so that excellent yields of, for example, neopentyl or cyclobutyl halides may be obtained in this way from the appropriate carboxylic acid precursors. This provides another alternative to the Hunsdiecker reaction (Chapter 6). The mechanism in this case is believed to involve ligand transfer of halide from Pb^{IV} to an alkyl radical (Scheme 9.7).

$$RCO_2H + Pb(OAc)_4 \rightleftharpoons CH_3CO_2H + RCO_2Pb(OAc)_3$$

$$RCO_2Pb^{IV}(OAc)_3 \longrightarrow R\cdot + CO_2 + Pb^{III}(OAc)_3$$

$$Cl^- + Pb(OAc)_4 \rightleftharpoons CH_3CO_2^- + ClPb(OAc)_3$$

$$ClPb^{IV}(OAc)_3 + R\cdot \longrightarrow RCl + Pb^{III}(OAc)_3$$

Scheme 9.7

In our discussion of the applications of radical chemistry in synthesis (Chapter 6), the isolation of organometallic cobalt compounds was mentioned, and oxidative elimination was there seen to be important. With other metals, e.g. mercury, relatively strong carbon–metal bonding occurs. In the case of non-transition metal organomagnesium compounds, whilst CIDNP and other data indicate the occurrence of radical mechanisms for some processes, there has been an interesting controversy (see the bibliography) regarding the possible importance of radical intermediates in these and a number of other reactions generally accepted as text-book examples of processes involving two-electron shifts.

In summary, to attempt to distil a vast and complex subject into a few paragraphs is impossible, but perhaps enough has been said to allow the interested reader to delve more deeply into the relevant literature – which, after all, is a principal goal of this book.

10

Radicals in biology

10.1 OXIDISING RADICALS

The majority of living organisms depend on molecular oxygen to sustain life, yet the oxygen molecule is a biradical and, as we have seen in our brief excursions into autoxidation chemistry in earlier chapters, interaction of organic substrates with oxygen can lead to a host of complex oxidation products. Not surprisingly then, there is now wide recognition that, whilst some oxidising radicals serve a useful physiological function in the healthy organism, in other circumstances they can be associated with molecular damage and hence with disease. They are also thought to contribute to the processes of ageing. Under normal circumstances there exists a fine balance between the essential biochemistry of these oxidising species and the range of defence mechanisms which have been evolved to inhibit oxidation damage, not least the presence of the antioxidant vitamins, C and E – of which more later.

The toxic effects of molecular oxygen, for example its lethal nature in very small quantities to some anaerobic bacteria, or its damaging effect on almost all forms of higher animals and plants at partial pressures above the normal 20% of atmospheric, have long been established, and were first interpreted in terms of molecular effects of oxygen-centred radicals in the mid-1950s. House-flies which have been surgically deprived of their wings, or confined in a small bottle, are especially long-lived. It has been argued that this is because, if they cannot fly, their mean respiration rate – and therefore their oxygen uptake – is greatly reduced! On the other hand, evidence that the life-expectancy of laboratory animals can be increased by a diet rich in synthetic antioxidants such as BHT (see Section 9.5) seems suspect.

Two rather obvious processes are available whereby molecular damage may be brought about by the intervention of oxygen-centred radicals. The first involves radical-chain autoxidation, to which polyunsaturated lipids containing 1,4-diene units are particularly susceptible (Section 9.4). To some extent, however, this is an important positive process, for, mediated by iron-dependent oxygenase enzymes, such chemistry leads to the important prostanoid and leukotriene hormones (see below).

1 cal = 4.184 J

The second process depends on the formation of hydroxyl radicals. These are so reactive and unselective that they will attack almost any organic molecule at, or very close to, the diffusion-limited rate. This implies the possibility of damage to proteins, DNA, lipids or carbohydrates with little scope for significant prevention by so-called hydroxyl-radical scavengers. Nature has an alternative approach wherein it strives to circumvent hydroxyl radical production. We shall begin by examining this.

A simplistic route to hydroxyl radicals in a biological oxidation would involve two-electron reduction of O_2 to hydrogen peroxide and subsequent catalysed decomposition in a Fenton-like reaction (Chapter 3). Is this possible? And how might hydrogen peroxide actually be formed? One-electron reduction of oxygen gives the superoxide radical anion, $O_2^{\cdot-}$, and in the course of normal biological processes this does happen to a modest extent. For example, in red blood cells the familiar role of haemoglobin is reversible complexation with O_2, but occasionally the complex dissociates, giving $O_2^{\cdot-}$ together with the iron(III) species methaemoglobin which is no longer capable of binding molecular oxygen. Likewise, normal oxidative metabolism occurs with transfer of four electrons to O_2 and production of two molecules of water, but again the process occasionally diverts with production of superoxide.[1] Superoxide has a pK_a of 4.8, so that at physiological pH a small fraction is protonated to form HOO·. The reaction between this and superoxide is extremely rapid (k = ca. 10^8 M^{-1} sec^{-1}) giving oxygen and hydrogen peroxide [Equation (1)]. The corresponding reaction between two superoxide molecules is much slower. If this source of hydrogen peroxide seems

$$HOO\cdot + O_2^{\cdot-} + H^+ \longrightarrow O_2 + H_2O_2 \qquad (1)$$

alarming, the presence of one of the superoxide dismutase enzymes (SODs) which *catalyse* the reaction might engender even greater concern! However, nature has devised ways of dealing both with hydrogen peroxide, and with any iron which may catalyse Fenton chemistry. Two enzyme types are available to destroy hydrogen peroxide. Catalase enzymes promote the transformation of two hydrogen peroxide molecules into two molecules of water and one of oxygen.[2] Peroxidases behave rather differently, depending on a hydrogen donor [DH_2 in Equation (2)] to deliver

$$H_2O_2 \xrightarrow[\text{peroxidase}]{DH_2} 2H_2O \qquad (2)$$

[1]Details of the one-electron transfers, which involve *inter alia* quinone radical anions, are beyond the scope of this chapter. The interested reader is referred to the bibliography at the end of the book.

[2]The oxygen is formed directly in its triplet state.

hydrogen. In mammalian cells the hydrogen donor is the tripeptide glutathione (**1**), two molecules of which are oxidised to the corresponding disulphide.[1] This reaction is coupled to a further enzyme-catalysed process which regenerates the glutathione. The redox active sites in (many) SODs, catalases, and glutathione peroxidase depend respectively on copper and zinc, an iron porphyrin, and selenium.

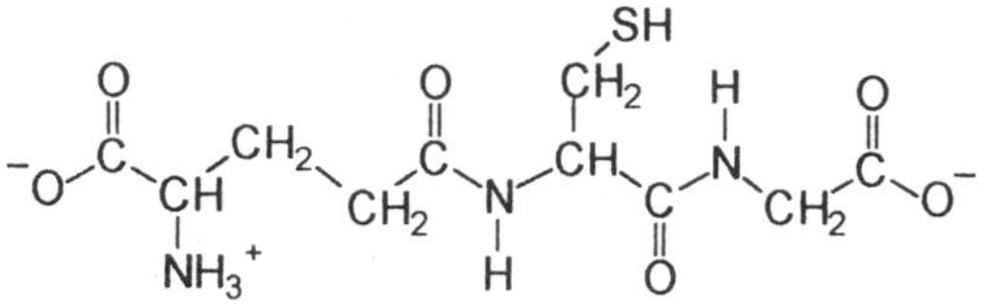

(**1**)

Not only are there enzymes present which effectively remove hydrogen peroxide. In normal cells any iron is in a form which cannot promote Fenton chemistry. Predominantly, it is complexed as Fe(III) to a glycoprotein called ferritin. There are, however, certain disease conditions in which this no longer holds true. The hereditary anaemia in humans known as thalassaemia, in which haemoglobin production is retarded, originally led to death in infancy, but it was found to be treatable by repeated blood transfusions. Unfortunately, the body is not very good at removing the iron that accumulates from this form of therapy, and this results in a condition known as "iron overload", in which some of the iron is present in a form uncomplexed by protective proteins and able to promote Fenton chemistry. The consequence again was premature death, usually in the late teens or early twenties, and often from heart failure, since the heart is one of the organs in which the iron tends to accumulate. For the Fenton chemistry to be truly catalytic, the resultant iron(III) must be reduced back to iron(II). A principal reducing agent is none other than superoxide. Since the early 1960s, the iron overload problem has been largely overcome by treating patients with efficient iron chelating drugs such as the tris-hydroxamic acid, desferrioxamine (**2**), isolated from a *Streptomyces* strain. These

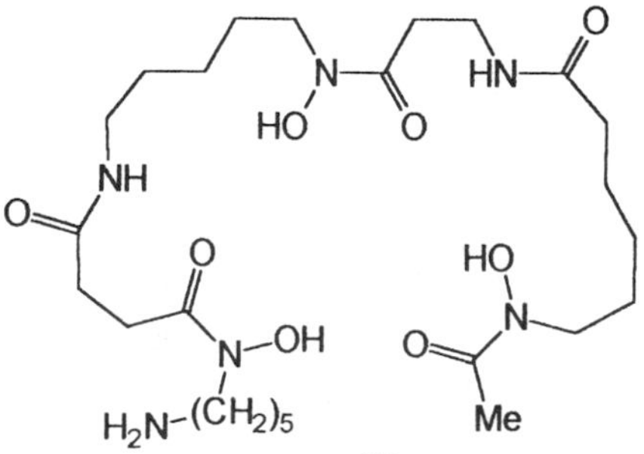

(**2**)

[1]Thiols are particularly susceptible to radical oxidation which commonly leads to disulphides ($2RSH \rightarrow RSSR$). The reaction is unsurprising; cf. the S–H bond strength of a thiol (Table 4.3).

1 cal = 4.184 J

so strongly bind the excess iron(III) that the reaction with superoxide is prevented. Furthermore, the complexes are capable of being excreted, so that iron concentrations in the body are maintained at reasonable levels.

Of course, there are other routes both to superoxide and to hydroxyl radicals which depend on extraneous influences. Exposure to ionising radiation is a case in point, where radiolytically produced electrons can be captured by molecular oxygen, or radiolysis of water can give hydroxyl directly. The consequent radiation sickness, be it a side effect of radiotherapy or the terrible aftermath of a nuclear accident, is well known.

There is now a wealth of examples of the damaging effects of hydroxyl radicals. An interesting one from veterinary medicine involves ocular damage in cattle treated against parasite infection with phenothiazine (**3**). This has been interpreted in terms of metabolic oxidation at sulphur and photolysis in the eye of the protonated sulphoxide. The products are hydroxyl and the rather stable phenothiazine cation radical. The latter can be re-oxidised to sulphoxide, making the process catalytic.

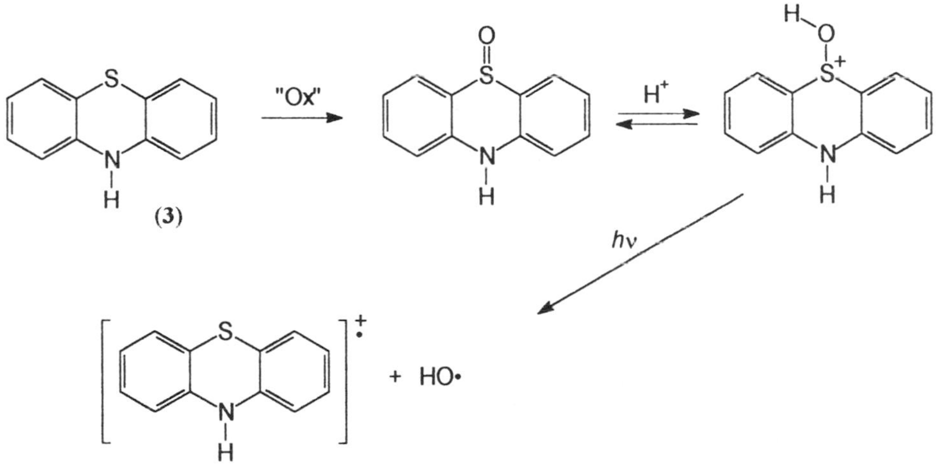

Once radicals are produced in the body, there is the possibility of autoxidation of polyunsaturated lipids. Protection against this is afforded especially by the lipid soluble phenolic antioxidant vitamin E (α-tocopherol; **4**). There is a synergism between this and the more abundant but water-soluble vitamin C (ascorbic acid; **5**), such that tocopherol is regenerated from its oxidised form at the water–lipid interface (Scheme 10.1; overleaf). The oxidised form of vitamin C is a highly delocalised radical anion. Because of its stability, this is often detectable in biological materials by ESR.

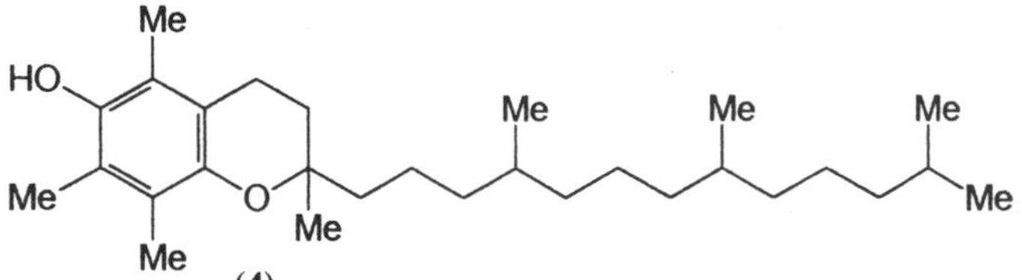

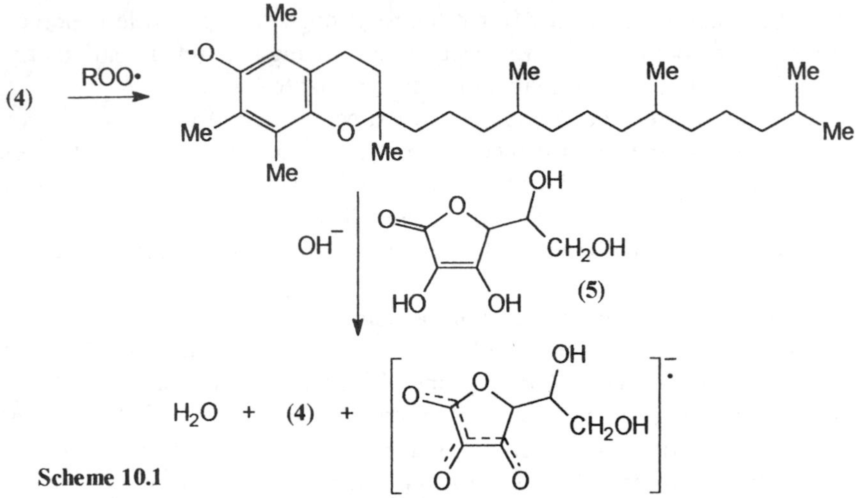

Scheme 10.1

Antioxidant properties are also attributed to selenium, an essential trace element, and to vitamin A. The key role of selenium is probably as the readily oxidised amino acid selenocysteine at the active site of glutathione peroxidase (see above); the precise role of vitamin A (retinol; **6**) seems less clear.

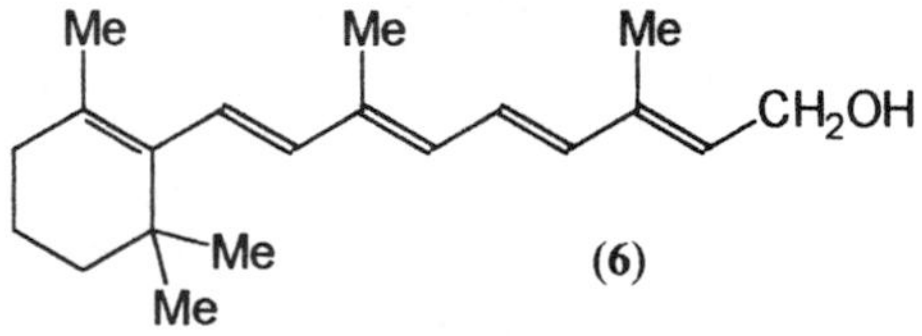

(6)

Mechanistic details almost invariably remain to be worked out, but the importance of radicals in disease has become increasingly clear during the past two decades. For example, whilst the onset of arthritis may not be induced by oxidising radicals, their presence in the tissue of inflamed joints is well established. Similarly, whilst radicals may not induce Parkinsonism, it has been suggested that continued high dosages of the antioxidant vitamins may delay the onset of the more advanced stages of the disease. A new dimension has been added to this area of research by the recent discovery that inherited motor neurone disease[1] is associated with a defect in the gene which normally codes for production of a copper/zinc superoxide dismutase.

One particularly well-documented situation relates to tissue damage following a heart attack. Blockage of the coronary arteries which supply heart muscle with oxygenated blood may be only transitory, and not itself cause cell death. However, serious injury may be caused when the blood starts flowing again – referred to as

[1]Known in the USA as familial amyotrophic lateral sclerosis.

1 cal = 4.184 J

"reperfusion damage". A plausible interpretation of this in some animals is in terms of an accumulation of both hypoxanthine (**7**) and the enzyme xanthine oxidase during the oxygen-deficient ("ischaemic") phase. When the oxygenated blood supply is restored, the accumulated enzyme catalyses rapid oxidation of the hypoxanthine. At the same time there is massive over-production of superoxide, leading to irreversible cell damage. Support for this interpretation comes from the minimisation of cell damage in animal experiments in which reperfusion is accompanied by administration of superoxide dismutase or desferrioxamine (**2**). However, xanthine oxidase is absent from human heart tissue.

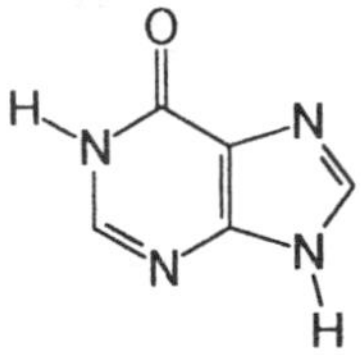

(**7**)

As newer results lead to more detailed information on the structures and modes of action of different enzymes, unexpected and surprising results on the molecular diversity between and within species are constantly coming to light. For example, as well as the Cu/Zn superoxide dismutases, enzymes with similar catalytic activity depending on iron or manganese are known. The manganese enzymes, which, like the copper/zinc enzymes, are found in man, have very different protein structures, and exhibit a different pH profile for their action. Recently it has been found that different individuals may have one of two variants of Mn-SOD. Both are protein tetramers incorporating four manganese atoms, but the replacement of isoleucine by threonine at one point in the protein chain of each quadrant of the enzyme results in a much more open structure for the molecule, and a significantly reduced catalytic activity. The intriguing suggestion has been made that evolution has not eliminated this less active variant because many pathogens are particularly sensitive to superoxide so that its presence imparts a greater resistance to disease. On the other hand, subjects possessing it may be particularly prone to some of the degenerative diseases of old age which may be induced by oxidative damage to cells.

An important biochemical pathway in the body involves the homologation and desaturation of linoleic acid (see p. 119) with the formation of the C_{20} tetraene-carboxylic acid, arachidonic acid (**8**). It is this which is the precursor of the prostaglandins, e.g. (**9**), as well as a wide variety of other potent bioactive species,

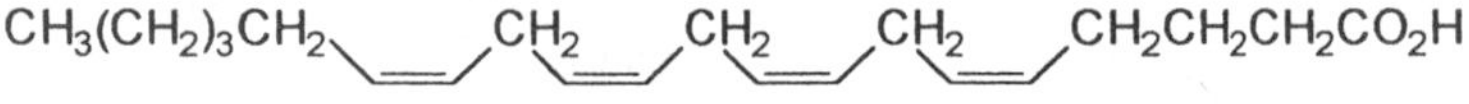

$CH_3(CH_2)_3CH_2$ ... CH_2 ... CH_2 ... CH_2 ... $CH_2CH_2CH_2CO_2H$

(**8**)

including the leukotrienes, e.g. **(10)**, prostacyclin **(11)**, and thromboxanes, e.g. **(12)**. The range of processes leading to these is sometimes referred to as the "arachidonic acid cascade" (which is activated *inter alia* by cell injury). We shall, however, confine our attention to the production of the prostaglandin skeleton, commencing

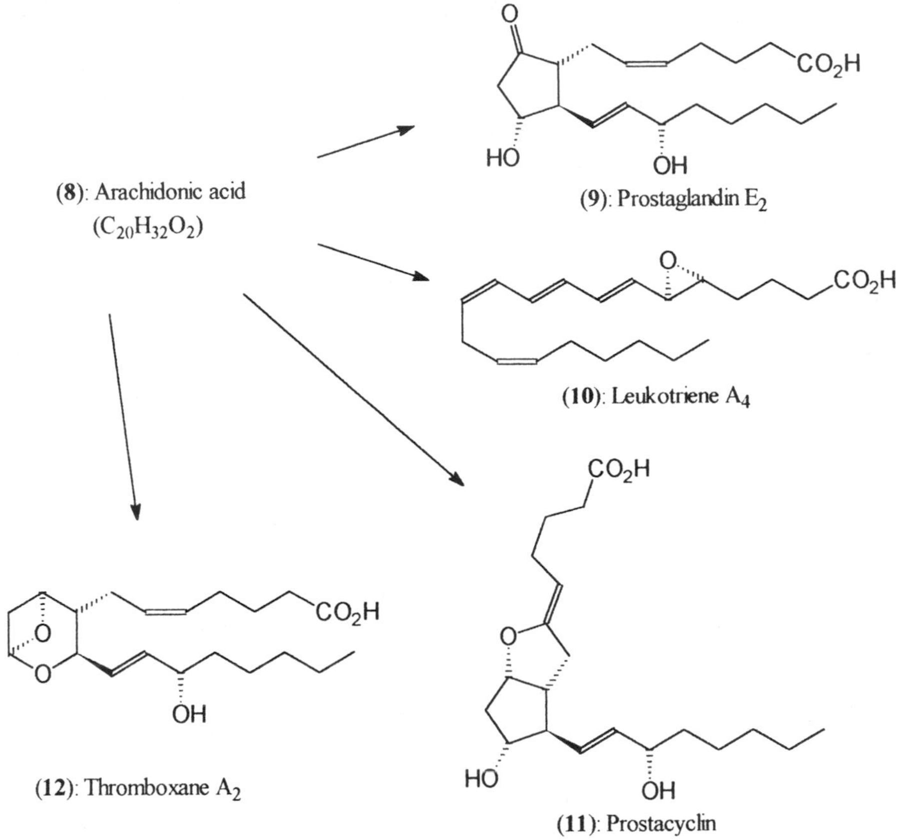

(8): Arachidonic acid ($C_{20}H_{32}O_2$)

(9): Prostaglandin E_2

(10): Leukotriene A_4

(12): Thromboxane A_2

(11): Prostacyclin

with arachidonic acid drawn in the "folded" representation **(8a)**. Hydrogen abstraction from the doubly allylic methylene at C-13 gives a conjugated dienyl radical which combines with molecular oxygen at C-11. Examination of the structure of the resulting peroxyl radical reveals that it is a "dioxahex-5-enyl radical"; like other hexenyl radicals, this cyclises to give a five-membered ring **(13)**. Structure **(13)** also incorporates a hexenyl moiety. A second cyclisation on to the conjugated diene gives an allylic radical which is intercepted by a second oxygen molecule. Hydrogen abstraction by the new peroxyl radical completes the sequence, with the production of the bicyclic hydroperoxide **(14)**, known as prostaglandin G_2. Further prostaglandins (which are oxygenated derivatives of a parent "prostanoic acid"; **15)** are obtained by metabolic modification of **(14)**.

1 cal = 4.184 J

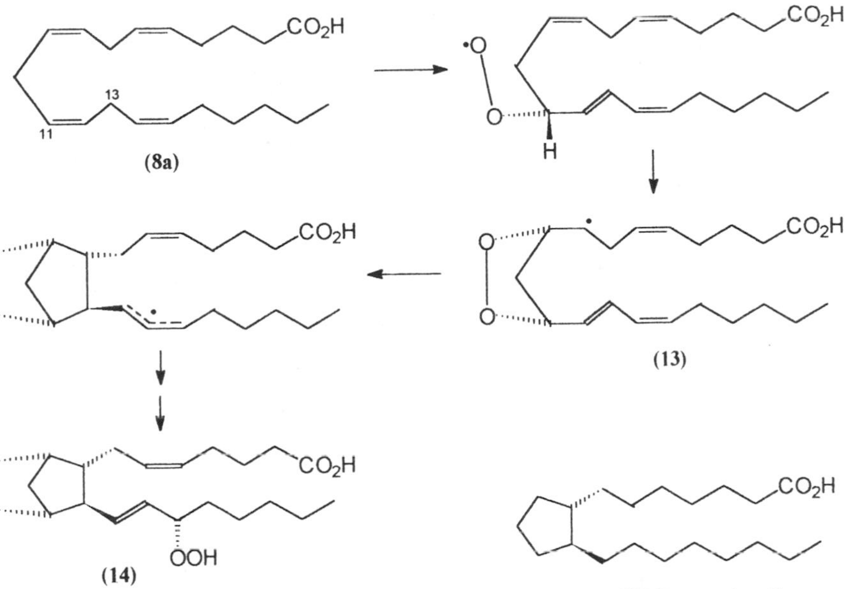

(8a)

(13)

(14)

(15) Prostanoic acid

Although prostaglandin-like biological activity has been found in products of an enzyme-free laboratory autoxidation of arachidonic acid, the importance of the cyclooxygenase enzyme in the natural systems is reflected in the apparently complete stereoselectivity of the transformations, e.g. **(8)** → **(14)**.

The many different products of the arachidonic acid cascade show a wide range of biological activity including dilatation of blood vessels and associated inflammation, smooth muscle contraction, and platelet aggregation.

10.2 ENZYME CATALYSIS MECHANISMS

Oxygen radicals have undoubtedly aroused high levels of interest in medical, biochemical and toxicological research, but from the standpoint of the molecular scientist some of the most profound recent work has been that which has been unravelling the radical character of a variety of enzyme-catalysed reaction mechanisms. One of the more thoroughly investigated is the group of enzymes collectively referred to as cytochrome P-450. These are found in both animal and plant tissues. Their active site incorporates an iron-porphyrin, and they function as "mono-oxygenases", i.e. they catalyse the incorporation of a single oxygen atom from O_2 into an organic substrate with simultaneous formation of a molecule of water – for which a reducing agent must also be present. The name P-450 derives from a characteristic absorption at 450 nm exhibited by a complex formed between the reduced form of the enzyme and carbon monoxide. Extensive studies of P-450 enzymes which catalyse the hydroxylation of alkanes have led to the conclusion that the oxidised [Fe(III)] form of the enzyme complexes the hydrocarbon (RH) and is then reduced (by NADPH catalysed by another enzyme called cytochrome P-450 reductase) to a new complex {[Fe(II)]-(RH)}. The iron then reversibly complexes molecular oxygen to form a species **(16)** which accepts a second electron (again from the cytochrome reductase) and two protons to release water and form the iron-oxo species **(17)**. The subsequent transfer of oxygen from iron to the bound

$$[\text{Fe}^{II}\text{–O}_2]\,[\text{RH}] \xrightarrow[e^-]{2H^+} [\text{Fe}^{IV}\text{=O}]^{+\bullet}\,[\text{RH}] + H_2O$$

(16) **(17)**

[(porphyrin disc) = porphyrin ligand]

hydrocarbon, with regeneration of the [Fe(III)] form of the enzyme, has some of the character of a radical process but other information had been interpreted in terms of direct "oxenoid" insertion into the C–H bond of the substrate [Equation (3)]. Thus retention of configuration at the site of oxygenation, and the finding of little or no primary kinetic isotope effect, both seemed inconsistent with a radical pathway. The

$$[\text{Fe}^{IV}\text{=O}\cdots\text{H–R}]^{+\bullet} \longrightarrow \text{Fe}^{III} + \text{ROH} \qquad (3)$$

Hypothetical "oxenoid" insertion into an aliphatic C–H bond

early isotope studies were based on *inter*molecular competition (e.g. between cyclohexane and cyclohexane-d_{12}). These had neglected the possibility that enzyme binding might be rate-limiting. When *intra*molecular competition was examined, large primary isotope effects ($k_H/k_D > 6$) were revealed. At the same time, examples of partial loss of configuration were encountered. This led to the conclusion that a hydrogen atom is removed from the substrate by the iron-oxo species and in a second, rapid step, referred to as "oxygen rebound", the hydroxyl ligand is transferred to the alkyl radical (Scheme 10.2). As a test of this hypothesis, and in an attempt to measure the rate of the oxygen rebound, the free-radical clock approach (Chapter 5) has been employed. For this the substrates chosen included *cis*- and *trans*-1,2-dimethylcyclopropane. Because the cyclopropyl C–H bonds are particularly strong (see footnote on p. 48), oxidation was correctly anticipated to occur at the methyl groups. But these substrates were selected because of the rapidity of the rearrangement of the corresponding cyclopropylmethyl radicals, the rate constants for which had previously been determined. Importantly, both of these

1 cal = 4.184 J

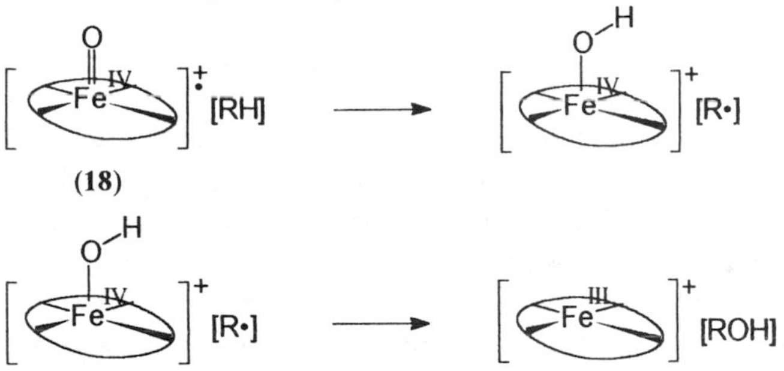

$$[Fe^{IV}=O]^{+\bullet}[RH] \longrightarrow [Fe^{IV}\text{-}OH]^{+}[R\cdot]$$

(18)

$$[Fe^{IV}\text{-}OH]^{+}[R\cdot] \longrightarrow [Fe^{III}]^{+}[ROH]$$

Scheme 10.2: The separate steps of the oxygen rebound mechanism.

radicals rearrange to a mixture of the two allylcarbinyl radicals (**18**) and (**19**). In the enzymatic oxidations, rearranged alcohols were formed in proportions corresponding, within experimental error, to the non-enzymatic results. Furthermore, since with each substrate some unrearranged alcohol was also formed, it was possible to use the known rates of opening of the cyclopropylcarbinyl radicals to estimate first-order rate constants for collapse of the [Fe(IV)-OH][R·] pairs. In both cases this was ca. 2×10^{10} sec^{-1}.

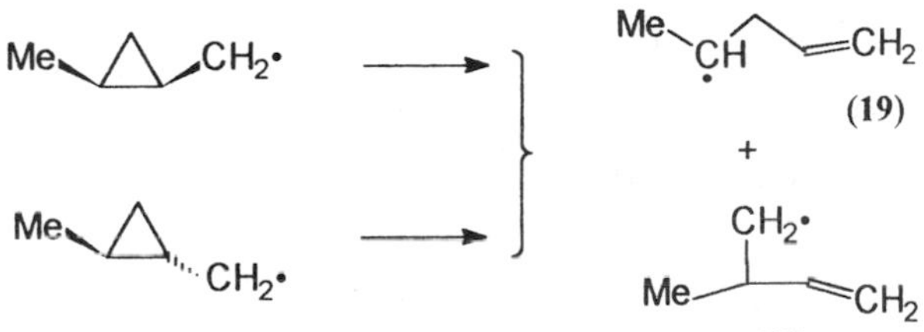

Me–cyclopropyl–$CH_2\cdot$ → Me–$\dot{C}H$–CH_2–CH=CH_2 (19)

+

Me–cyclopropyl–$CH_2\cdot$ → $CH_2\cdot$–C(Me)=CH_2 (19)

Not all of the alkylcyclopropanes investigated in this study underwent oxidation and rearrangement as described above, and it was concluded that some substrates were simply too large to enter the active site. This type of investigation has greatly illuminated the manner in which our understanding of free-radical behaviour in a pure solvent has to be modified in any attempt to comprehend what may occur in the constrained environment of the active site of an enzyme.

Other enzymes, such as horseradish peroxidase, and the non-haem iron enzymes typified by methane mono-oxygenase (which catalyses methanol formation) and isopenicillin N synthase [which catalyses cyclisation of a tripeptide to the penicillin nucleus; Equation (4)], are believed to operate through related iron-oxo species.

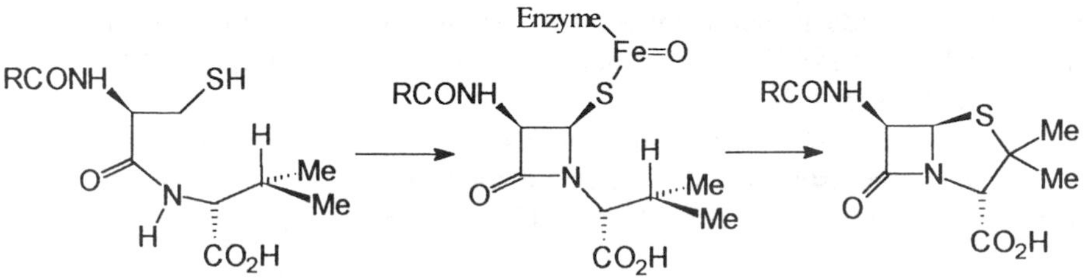

RCONH, SH, H, Me, Me, O, N, H, CO_2H → Enzyme, Fe=O, RCONH, S, H, O, N, Me, Me, CO_2H → RCONH, S, Me, Me, O, N, CO_2H (4)

R = L-$HOCOCH(NH_2)CH_2CH_2CH_2$-

In the reaction promoted by the isopenicillin synthase, an acyclic tripeptide is first converted to β-lactam. The subsequent hydrogen-abstraction/ligand transfer has been formulated as involving the *S*-ferryl intermediate shown in Equation (4), but there is not necessarily an implication of synchronicity in this second step, and indeed suitably modified substrates give products characteristic of alkyl-radical rearrangement, e.g. **(20)** → **(23)**. However, further studies with regio- and stereo-specifically deuterated derivatives of **(20)** have led to the assertion that, following initial β-lactam formation **(21)**, the rearrangement most probably involves a concerted oxenoid type of insertion accompanied by cyclopropyl ring-opening to form the iron heterocycle **(22)**, but that radical intervention does then occur in the subsequent conversion of **(22)** into the seven-membered ring of **(23)**. This was based on the discovery, from the labelling experiments, that it is exclusively the a–b bond in **(20)** which opens, but that the configuration at b is scrambled. The

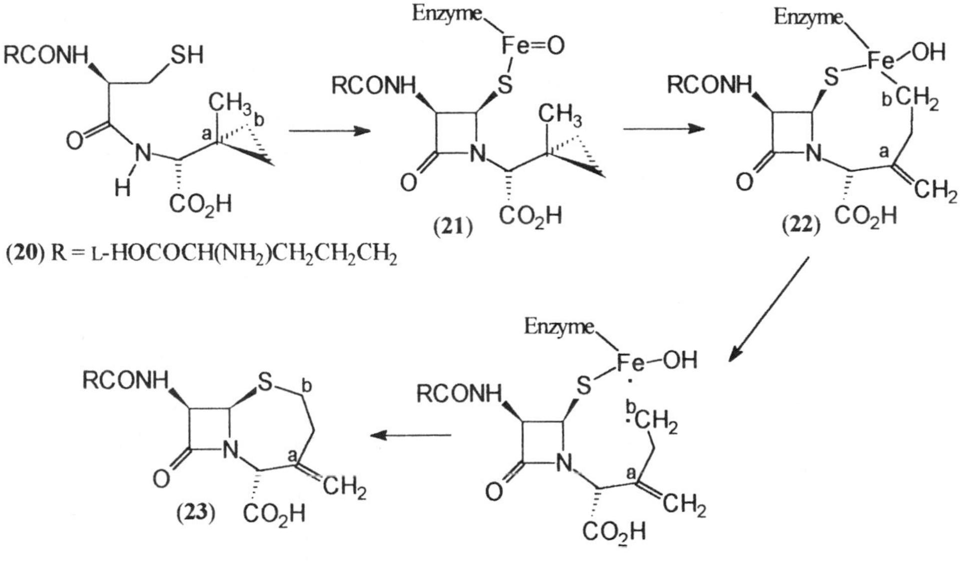

(20) R = L-$HOCOCH(NH_2)CH_2CH_2CH_2$ **(21)** **(22)** **(23)**

alternative rationalisation of this selectivity, namely that the opening of a cyclopropylcarbinyl radical could occur with such high regioselection *governed by the constraining influence of the catalytic site in the enzyme*, was considered less likely. Apparently, therefore, there is a significant distinction between the isopenicillin N synthase and the cytochrome P-450. Conceivably, this distinction may be found to reside in some inherent difference in behaviour of haem iron-oxygenases such as the P-450 enzymes and their non-haem counterparts represented by the isopenicillin synthase.

Another area which has commanded intensive scrutiny, and in which radical-mediated enzyme chemistry is now well established, involves the various processes dependent upon coenzyme B_{12} (Fig. **10.1**). These all reveal apparent 1,2-shifts of

1 cal = 4.184 J

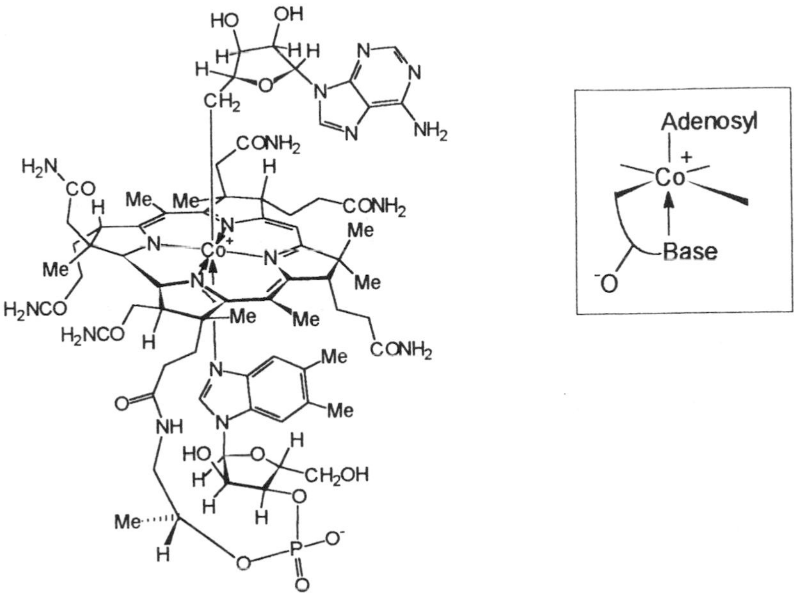

Fig. 10.1: The full structure of coenzyme B_{12} (adenosylcobalamin) and a simplified representation.

hydrogen offset by 2,1-shifts of an electronegative group. Representative examples are methylmalonyl-coenzyme A mutase, Equation (5), and various diol dehydratases, e.g. Equation (6). A recent mechanistic proposal for the latter system is embodied

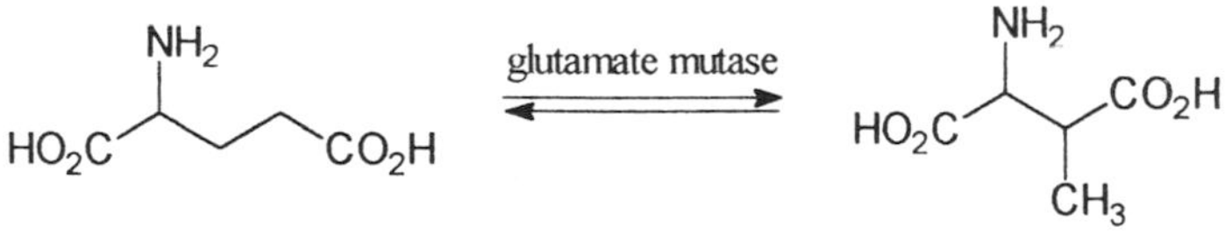

$$HO_2C\text{-}CH_2\text{-}CH_2\text{-}CH(NH_2)\text{-}CO_2H \xrightleftharpoons{\text{glutamate mutase}} HO_2C\text{-}CH(NH_2)\text{-}CH(CH_3)\text{-}CO_2H \quad (5)$$

$$HOCH_2\text{—}CH_2OH \xrightleftharpoons{\text{diol dehydratase}} CH_3\text{—}CHO + H_2O \quad (6)$$

in Scheme 10.3 (overleaf), in which the adenosyl–cobalt bond of the coenzyme cleaves (step i), and the resulting adenosyl radical abstracts hydrogen from an enzyme-bound site (ii); this gives an enzyme-bound radical which in turn removes hydrogen from the diol *in a propagating step of a radical-chain cycle* (iii). The intermediate 1,2-dihydroxyethyl radical is bound to acidic and basic sites on the enzyme in such a manner that hydroxyl migration is facilitated (see below). The cycle is completed by the return of the hydrogen atom from the radical site on the enzyme (iv), followed by dehydration of the resulting 1,1-diol and site exchange between it and a second molecule of 1,2-diol. Chain termination occurs if the enzyme radical abstracts hydrogen from adenosine and the adenosyl–cobalt bond is reformed. The wealth of chemical, stereochemical and other evidence that appears to require intervention of a protein-bound radical or initiation by fission of the weak

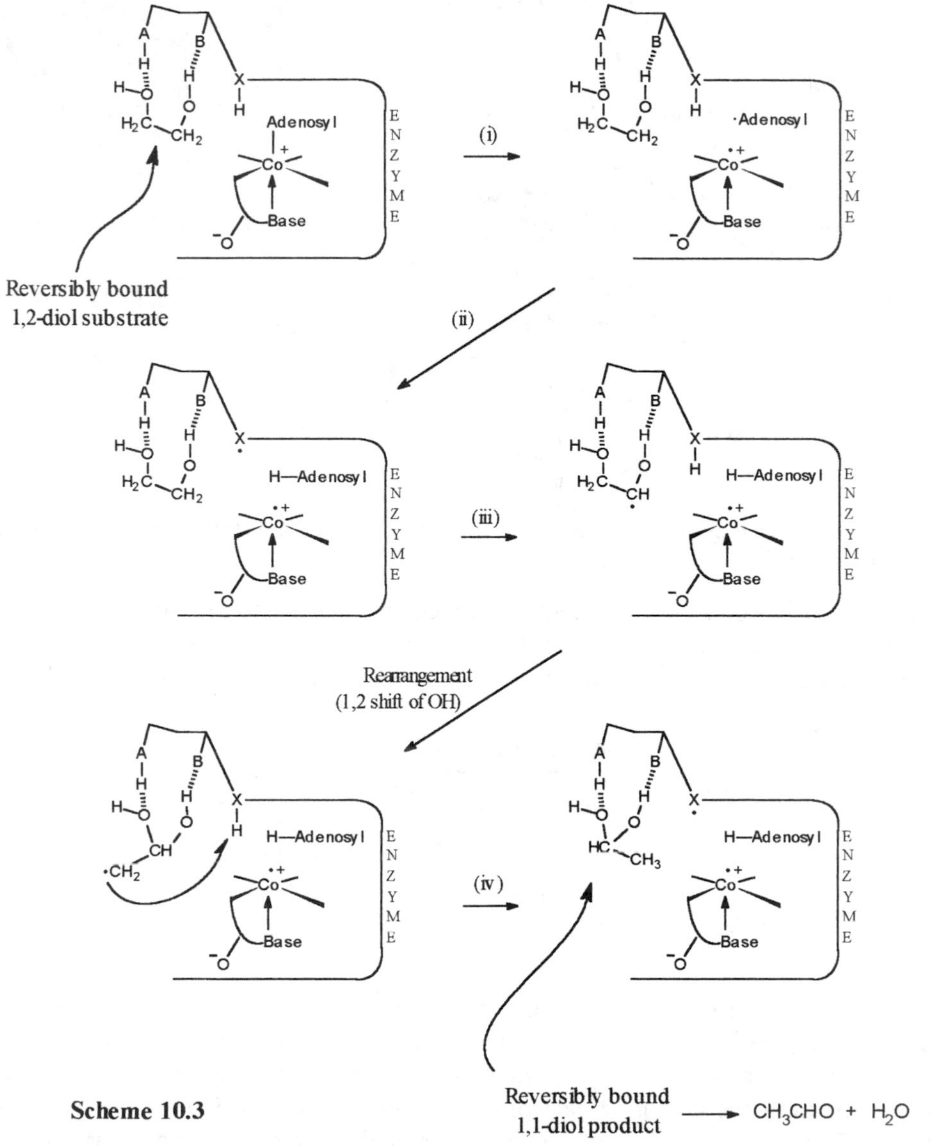
Reversibly bound
1,2-diol substrate
(i)
(ii)
(iii)
Rearrangement
(1,2 shift of OH)
(iv)
ENZYME
Adenosyl
·Adenosyl
H—Adenosyl
Base
Reversibly bound
1,1-diol product
CH3CHO + H2O

Scheme 10.3

carbon–cobalt bond cannot be addressed here. However, one significant aspect of this work has an important precedent in simple radical chemistry, namely the facilitation of hydroxyl-group migration in the intermediate radical by protonation.

1 cal = 4.184 J

Early ESR studies of reactive radicals in aqueous solutions, using the flow-through technique which was outlined in Chapter 5, had established that the principal radical detected following hydrogen abstraction from ethylene glycol (**24**) by hydroxyl radicals is dependent on pH. As the acidity is increased, signals due to (**25**) are replaced by those from (**26**). The interpretation of these observations was in terms of Equation (7). This loss of water, if followed by hydration of the formyl group, amounts to an acid-catalysed migration of hydroxyl.

$$\underset{(\mathbf{24})}{HOCH_2CH_2OH} \xrightarrow{HO\cdot} \underset{(\mathbf{25})}{HOCH_2\dot{C}HOH} \underset{-H^+}{\overset{H^+}{\rightleftharpoons}} H_2\overset{+}{O}CH_2\dot{C}HOH \xrightarrow{-H^+} \underset{(\mathbf{26})}{\cdot CH_2CHO} \quad (7)$$

$$\left[\cdot CH_2CHO \underset{H_2O}{\overset{-H_2O}{\rightleftharpoons}} \cdot CH_2{-}CH(OH)_2 \right]$$

It is clear that dehydration of (**25**) may also be catalysed by base [Equation (8)]; in the enzyme, catalytic sites of both kinds are involved in binding the substrate.

$$\underset{(\mathbf{25})}{HOCH_2\dot{C}HOH} \xrightarrow{HO^-} HOCH_2\dot{C}HO^- \xrightarrow{-HO^-} \underset{(\mathbf{26})}{\cdot CH_2CHO} \quad (8)$$

The ribonucleotide reductase enzymes, which are present in many cells, and which catalyse the replacement of a hydroxyl group from ribonucleoside diphosphates by hydrogen, have also been the subject of extensive mechanistic investigation. By no means all of their mysteries have been revealed, but it is clear that a common feature is the ability to generate an organic radical. Some ribonucleotide reductases have be shown spectroscopically to contain a phenolic radical, in the form of an oxidised tyrosine residue embedded in the protein chain. This appears to be crucial for initiating an electron-transfer relay to the active site at which another protein radical is derived from XH (see Scheme 10.4, overleaf). This is probably a thiyl radical, i.e. X = S in the scheme, which then abstracts hydrogen from C-3' of the ribose moiety of the substrate. Two further thiol groups are oxidised to disulphide whilst the oxygen of the hydroxyl at C-2' of the ribose is reduced to water. A highly abbreviated picture of the essential features of this process is presented in the scheme.

Yet another protein radical is encountered in reactions involving pyruvate formate lyase (PFL). This enzyme catalyses a key step in bacterial fermentation of glucose in which pyruvate is cleaved thiolytically to give formate and acetylcoenzyme A [represented by Equation (9)]; it has been established that the reaction depends on a dehydrogenation of a glycine residue in the enzyme to a glycyl radical (**27**). Since the radical centre is located on carbon lying between electron-acceptor carbonyl and -donor nitrogen, it is interesting to speculate on how far this affords a biological example of captodative stabilisation (Section 9.3). This recent discovery has led to the search for the formation of similar species in other enzyme systems.

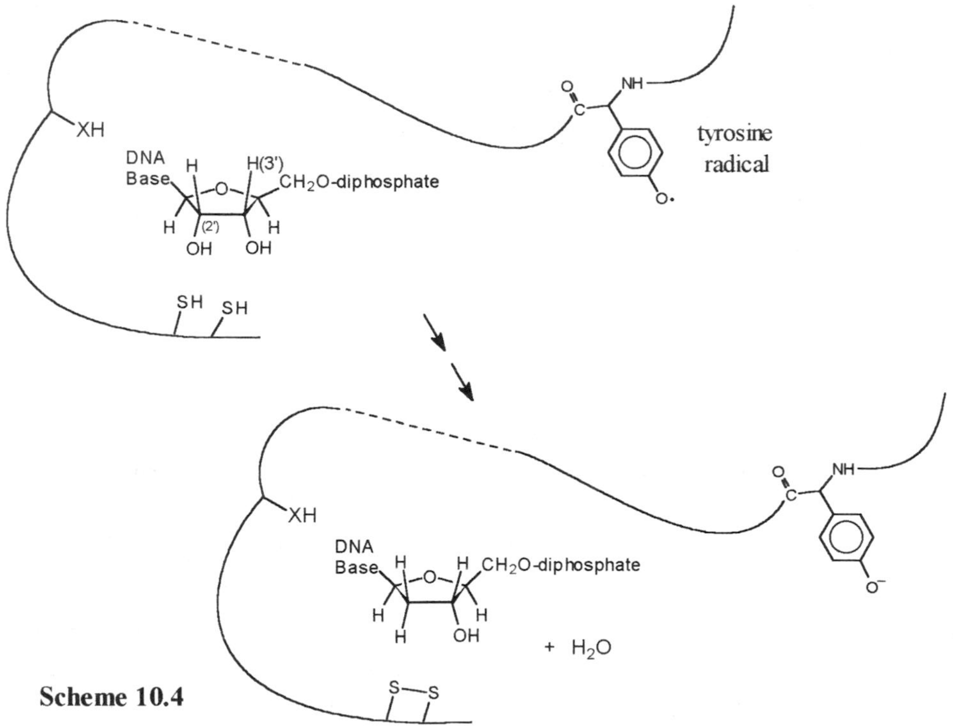

XH
DNA Base
H
H(3')
CH_2O-diphosphate
H (2') OH OH H
SH SH
O C NH
tyrosine radical
O·

XH
DNA Base
H H CH_2O-diphosphate
H H OH H
+ H_2O
S—S
O C NH
O^-

Scheme 10.4

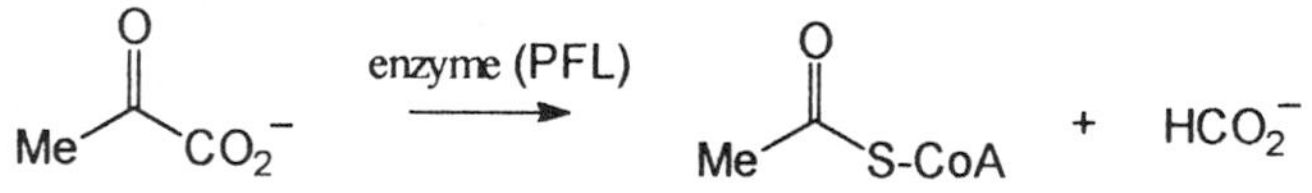

$$\mathrm{MeCOCO_2^-} \xrightarrow{\text{enzyme (PFL)}} \mathrm{MeCO\text{-}S\text{-}CoA} + \mathrm{HCO_2^-} \qquad (9)$$

$$\mathrm{\cdots NH\text{-}CO\text{-}CH_2\text{-}NH\text{-}CO\cdots} \xrightarrow{X\cdot} \mathrm{\cdots NH\text{-}CO\text{-}\dot{C}H\text{-}NH\text{-}CO\cdots}$$

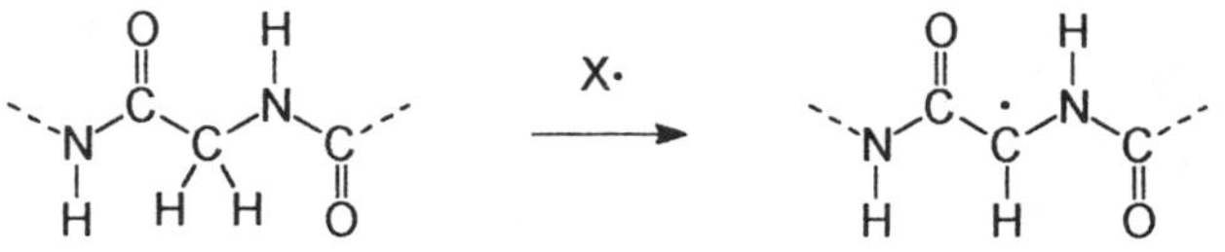

(27)

1 cal = 4.184 J

10.3 SPIN TRAPPING AND SPIN LABELLING

Of the several ESR techniques which have found applications to biological systems, we shall mention briefly only two. Spin trapping, introduced in Chapter 5, has been widely employed to study biological oxidations. The nitrone traps, e.g. PBN and DMPO, have been used most frequently since these give the longest-lived spin adducts with oxygen-centred radicals. But, as indicated earlier, there are many pitfalls for the unwary, including difficulty in interpreting the very simple spectra usually obtained from nitrones, the formation of secondary nitroxides, and the formation of nitroxides by routes other than radical trapping. A preliminary report that the water-soluble nitrosoaromatic trap **(28)** efficiently and rather selectively scavenges superoxide prompted considerable interest in the biomedical community

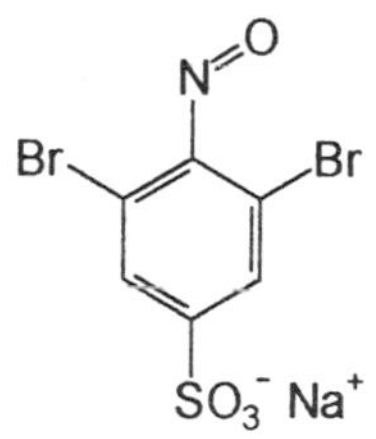

N=O, Br, Br, SO_3^- Na^+

(28)

a few years ago and led to the initiation of a number of new ESR investigations. The immediate goal of each of these was in vain, however, since it was soon shown that the supposed superoxide adduct was apparently an artefact of the solvent system used in the original work. Whilst these comments are issued as a warning, there is no question that in some biological investigations the use of spin traps has been qualitatively invaluable. Some semi-quantitative data obtained by spin trapping also seem to tie in with results obtained by alternative procedures, and whether or not there is even order of magnitude accuracy in a report that 10^{14} radicals may be spin trapped from a single puff of cigarette smoke, the alarm bell tolls loud!

As a general rule, however, it is a wise precaution to corroborate any spin-trapping results by using suitable complementary procedures.

The title "spin trapping" probably was coined by analogy with a slightly earlier application of ESR which also depends on nitroxides and had been named "spin labelling". In this, some species of biochemical interest is tagged with a stable nitroxide radical which then performs as a "reporter group". The object of this approach is that only the reporter group will be active in some spectroscopic monitoring procedure (in this case ESR) whilst the rest of the system will be "silent". Other reporter groups include fluorescent labels and, of especial interest because the system suffers negligible structural perturbation, specific isotope labels such as deuterium, the latter being used in conjunction with the appropriate, e.g. deuterium, NMR technique.

Nitroxide spin labels with various kinds of functionality are available which have been tailored to react with specific sites in molecules of interest. One example is **(29)**, in which the labile iodoacetamide group is readily attached to nucleophilic sites in an enzyme or other substrate. The ESR spectrum then "reports" on factors like the mobility of the labelled site, and the polarity of its immediate environment. Remembering (Chapter 5) that the nitroxide structure can be represented as a hybrid of the two canonical forms **(30a)** and **(30b)**, it is not particularly surprising that the balance between these is perturbed by the polarity of the immediate surroundings, and that this is manifest in the magnitude of the nitrogen hyperfine interaction, a_N. Thus in hexane a_N is usually slightly less than 15 gauss whereas in water it rises to almost 17 gauss and goes higher still in solutions of high ionic strength. Indeed, the nitrogen hyperfine splitting of di-t-butyl nitroxide has been advocated as the basis of a solvent-polarity scale, giving excellent correlations with some known solvent effects.

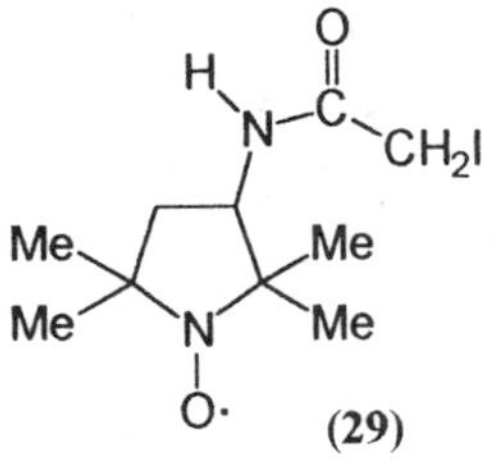

(29)

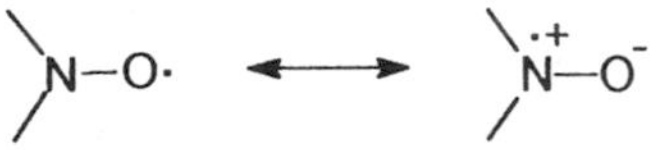

(30a) **(30b)**

The approach to mobility depends on the magnetic anisotropy of the nitroxide moiety. This anisotropy is nicely illustrated by spectra obtained using a dilute *solid* solution of di-t-butyl nitroxide in a single crystal of 2,2,4,4,-tetramethylcyclobutanone. The nitroxide fits into lattice sites in the crystal, so that all the nitroxide molecules are orientated in the same direction. Examination of this crystal in the spectrometer reveals three sharp lines due to the nitrogen hyperfine interaction, but the positions of, and spacing between these depends on the orientation of the crystal, with limiting spectral parameters corresponding to orientation of the magnetic field along the x, y, and z axes defined by the nitroxide structure (Fig. **10.2**).[1] In mobile solvents of low viscosity, rapid tumbling averages the anisotropy effects to give the familiar three-line solution spectrum. But if the tumbling is restricted in some way, for instance by use of a viscous solvent or by rigid attachment of the nitroxide to a macromolecule whose tumbling rate will be

[1]This experiment cannot be duplicated on a crystal of pure nitroxide, since the close contact between the paramagnetic molecules results in rapid spin exchange (Chapter 5) and a single line spectrum is observed.

1 cal = 4.184 J

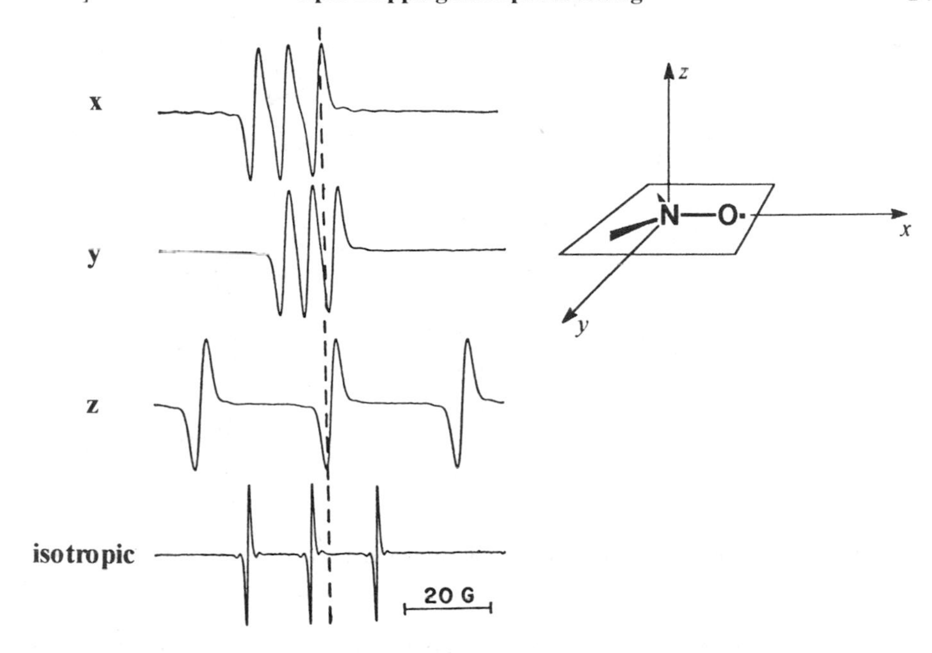

Fig. 10.2: ESR spectra of di-t-butyl nitroxide. The top three traces were recorded using a single crystal of 2,2,4,4-tetramethylcyclobutanone doped with the nitroxide (see text). The magnetic field was in each case aligned with one of the molecular axes of the radical (see inset). The bottom trace was obtained using a dilute (liquid) solution in di-t-butyl ketone. The dashed line represents a magnetic field marker from which it can be seen that not only a_N but also the g-value varies with the orientation of the sample. (Reproduced with permission from O.H. Griffith and A.S. Waggoner, *Accounts Chem. Res.*, **2**, 17 (1969)).

slow, effects of the anisotropy start to show up in a broadening of the spectral lines. Examination of Fig. **10.2** reveals that the limiting high-field lines (to the right of the spectra) are spaced furthest apart. Therefore when the spectra are incompletely averaged the most pronounced broadening shows up in the high-field component of the triplet. Less broadening is evident in the low-field component, whilst the central line remains reasonably sharp. Detailed analysis of these spectral features allows quantitative estimation of the mobility of the nitroxide probe. A computer simulation of a nitroxide spectrum exhibiting anisotropic broadening is traced in Fig. **10.3.** Spectral broadening of the kind described here reflects molecular motion on an approximately nanosecond time scale. More recently, techniques have been developed whereby the physics of signal saturation lends itself to probing much slower molecular motions of nitroxide-labelled species, such as translation and rotation of proteins and other macromolecules. These procedures have given a new impetus to spin-labelling experiments.

One practical, if regrettable, application of the spin-labelling technique was developed at the time of the Vietnam war, and was used as a rapid screening procedure for the abuse of hard drugs by US servicemen. It depended upon a spin-

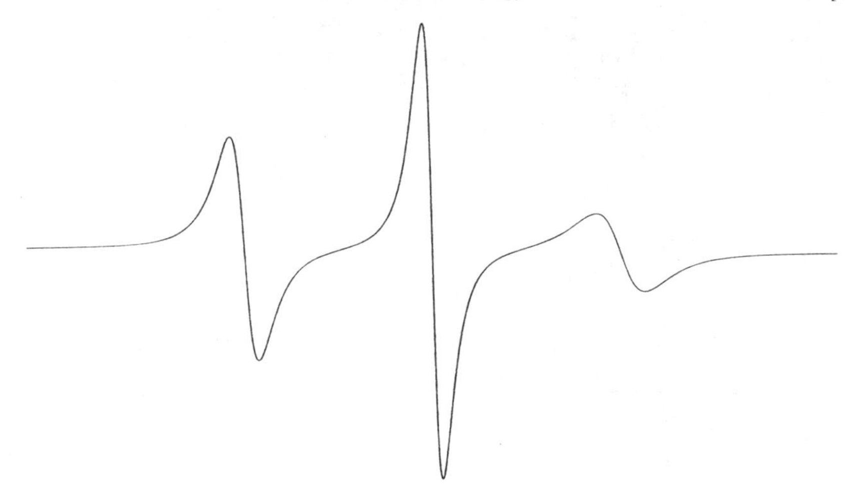

Fig. 10.3: Computer-generated spectrum of a nitroxide that might be observed using a solvent of high viscosity.

label immunoassay procedure based on the production of a morphine antibody complexed with the spin-labelled morphine derivative (**31**).[1] This nitroxide–protein complex, present at low concentration, tumbles very slowly so that the spectral lines are so broad as to be undetectable. However, when admixed with body fluids contaminated with the drug or a related metabolite, some of the labelled normorphine is displaced from its binding site. The small molecule liberated from the antigen tumbles rapidly, and since the same integrated signal intensity resides now

(31)

under relatively sharp lines, these are more easily detected against the background noise of the spectrum. In fact reasonable *quantitative* accuracy was reported.

Although a detailed discussion of ESR instrumentation, and of factors affecting the appearance of an ESR signal (instrument settings, electron exchange, small

[1]Readily prepared from normorphine and (**29**).

1 cal = 4.184 J

unresolved hyperfine interactions, etc.), is beyond the scope of this text, further insight into the consequences of varying line-width, evident in Fig. **10.3**, is provided in Fig. **10.4**. Two ESR lines of equal intensity are simulated, using identical line *shapes* (in this case Lorentzian), but varying the line *width* by a factor of three. A consequence of the first derivative format is that the peak *heights* differ by a factor in excess of eight!

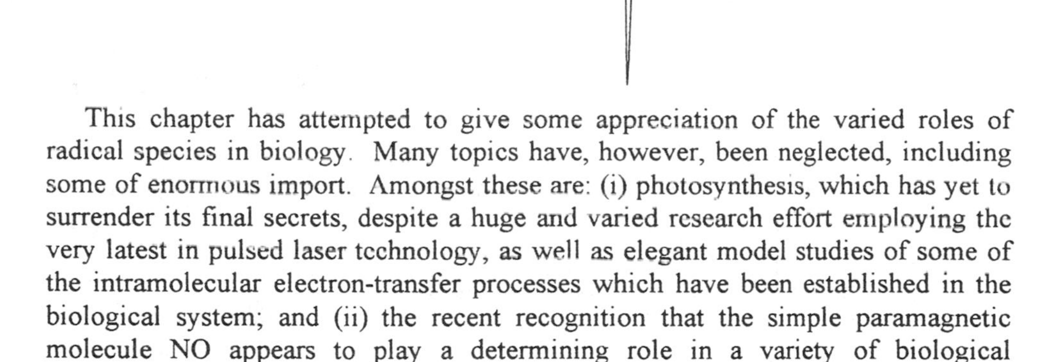

Fig. 10.4: Computer-generated spectrum showing two peaks of equal signal strength – see discussion in text.

This chapter has attempted to give some appreciation of the varied roles of radical species in biology. Many topics have, however, been neglected, including some of enormous import. Amongst these are: (i) photosynthesis, which has yet to surrender its final secrets, despite a huge and varied research effort employing the very latest in pulsed laser technology, as well as elegant model studies of some of the intramolecular electron-transfer processes which have been established in the biological system; and (ii) the recent recognition that the simple paramagnetic molecule NO appears to play a determining role in a variety of biological processes.[1]

10.4 THE ENEDIYNE ANTIBIOTICS

This text has also neglected biradicals (including carbenes and photo-generated excited states), with their interesting behaviour resulting *inter alia* from electron-spin correlation effects within single molecular entities. Therefore the concluding paragraphs of the book may seem an odd place to dwell on such species for the first time. However, some of the most potent cytotoxic agents to have been discovered in nature owe their toxicity to the very high radical reactivity of some rather unusual biradical species.

[1]Many possible roles for NO may be conjectured, including formation of metal complexes. One role which is of more immediate relevance to the discussion in this chapter is the rapid reaction with superoxide to form the peroxynitrite anion, $O{=}N{-}O{-}O^{-}$. Peroxynitrite (or the unstable peroxynitrous acid, HOONO) is likely to react as an oxidising species; whether or not it may constitute a source of hydroxyl radicals remains unclear.

We shall begin this extraordinary story by reviewing two chemical curiosities of the 1970s. The first is a peculiarly bizarre way of generating phenyl radicals [Equation (10)]. The second, a related pyrolytic reaction in which the 1,4-phenylenediyl (1,4-dehydrobenzene; **32**) was generated from hex-3-en-1,5-diyne, is now referred to after its discoverer as the Bergmann cyclisation; benzenoid 1,4-biradicals like (**32**) differ from 1,2-dehydrobenzene (i.e. benzyne), by showing high radical-like reactivity similar to that exhibited by phenyl monoradicals.[1]

CH·

(10)

(11)

(32)

In the late 1980s, no fewer than four new classes of naturally occurring antibiotics were discovered which incorporated enediyne or closely related groupings. What was special about these natural products was that they also incorporated further functionality, referred to as the "trigger", which, when activated, draws the ends of the enediyne unit close together, so that the Bergmann cyclisation ensues *at ambient temperature.*

It seems that these antibiotics associate with the cell's DNA in such a way that the trigger is released, the biradical is generated, and rapid reaction with the DNA then causes irreversible damage, apparently by strand breaking. The example of a calicheamycin (**33**) is illustrated in Scheme 10.4. In this case the trigger is released by nucleophilic attack on the unusual trisulphide group. At the time of writing, the mechanisms whereby radical attack on DNA may promote strand breaking are the subject of active investigation.

[1]In the early experiments, evidence for the formation of the 1,4-phenylenediyl was adduced from the production, *inter alia*, of 1,4-dideuterobenzene when a deuterated hydrogen donor was present. More recently, it has been found that, in the absence of hydrogen donor, the biradical will polymerise to form poly-*p*-phenylene, a polymer with remarkable thermal stability as well as other unusual solid-state properties.

1 cal = 4.184 J

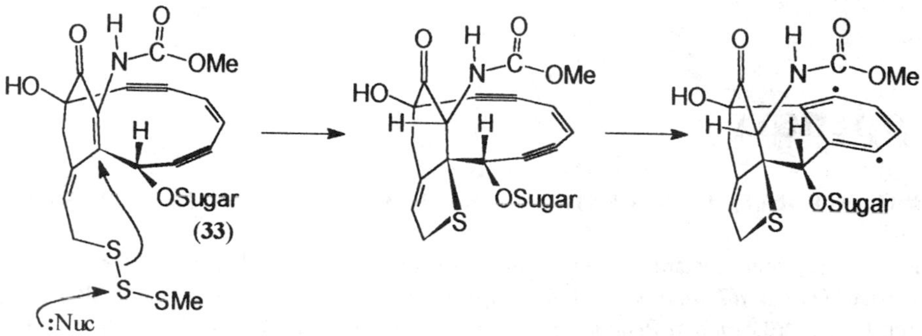

O
H
O
N—C—OMe
HO
H
OSugar
(33)
S
S—SMe
:Nuc
O
H
O
N—C—OMe
HO
H
H
OSugar
S
O
H
O
N—C—OMe
HO
H
H
OSugar
S

Scheme 10.4

Appendix

ABSOLUTE RATE CONSTANTS OF UNIT STEPS IN RADICAL CHEMISTRY

Many of the rate constants selected for inclusion in these tables have been taken from *Numerical Data and Functional Relationships in Science and Technology*, Llandolt-Bornstein, Group II, Vol. 13, *Radical Reaction Rates in Liquids*, Springer, Berlin, 1984. No error limits are quoted, but the comments regarding accuracy in Section 9.2, "Radical kinetics and thermodynamics", should be borne in mind.

Footnotes to the tables are collected at the end of this Appendix.

Section 1: Radical–radical reactions

Reacting radicals	Solvent	Temperature (K)	Rate constant (M^{-1} sec^{-1})
Me· + Me·	H_2O	298	3×10^9 ($2k_t$)
t-Bu· + t-Bu·	heptane	298	8.5×10^9 ($2k_t$)
$PhCH_2$· + $PhCH_2$·	H_2O	298	8×10^9 ($2k_t$)
t-BuOO· + t-BuOO·	benzene	295	4×10^2 ($2k_t$)[a]
Me_2CHOO· + Me_2CHOO·	Cl_2CF_2	229	8×10^4 ($2k_t$)[a]
p-$MeOC_6H_5O$· + *p*-$MeOC_6H_5O$· (→ *o*-coupled dimer)[b]	benzene	303	3×10^9 ($2k_t$)
2 Me—(2,6-di-Bu-t-phenyl)—O· (Bu-t, Bu-t)	toluene	293	4.5×10^7 ($2k_t$)
n-alkyl· + (2,2,6,6-tetramethylpiperidine N—O·)	isooctane	293	ca. 1×10^9 [c]

Section 2: Radical–molecule reactions

a: S_H2 Processes

Reacting species	Solvent	Temperature (K)	Rate constant (M^{-1} sec^{-1})
$Cl\cdot + RCH_3$	alkane	313	0.7×10^9 (per H)
$Cl\cdot + c\text{-}C_5H_{10}$	$c\text{-}C_5H_{10}$	313	4.7×10^8 (per H)[d]
$Cl\cdot + c\text{-}C_5H_{10}$	benzene	313	4.3×10^6 (per H)[e]
$Br\cdot + RCH_3$	alkane	313	ca. 20 (per H)
$t\text{-}BuO\cdot + RCH_3$	alkane	313	ca. 10^3 (per H)
$t\text{-}BuO\cdot + c\text{-}C_5H_{10}$	benzene	302	9×10^4 (per H)
$CH_3\cdot + RCH_3$	gas-phase	338	ca. 10 (per H)
$HO\cdot + RCH_3$	alkane	300	1×10^9 (per H)[d]
$Br\cdot + PhCH_2\text{-}H$	toluene	313	2×10^5 (per H)
$CH_3\cdot + Bu_3SnH$	cyclohexane	298	5.8×10^6
$Et\cdot + Bu_3SnH$	isooctane	300	2.3×10^6
$CH_3\cdot + MeSH$	H_2O	293	7.4×10^7
$PhCH_2\cdot + MeSH$	benzene	296	3.5×10^4
n-alkyl$\cdot$ + PhSH	nonane	298	1×10^8
$Ph_3C\cdot + PhSH$	toluene	315	15
n-alkyl$\cdot$ + PhSeH	THF	293	2×10^9 [f]
n-alkyl$\cdot$ + PhSeSePh	benzene	353	5×10^7
n-alkyl$\cdot$ + $BrCCl_3$	isooctane	353	1×10^8
cyclohexyl$\cdot$ + I_2	toluene	295	7×10^9
$Ph\cdot + CCl_4$	CCl_4	318	1×10^6 (per Cl)
$Et_3Si\cdot + Me_3CCl$	Et_3SiH	300	1.1×10^9
$Bu_3Sn\cdot + Me_3CCl$	cyclohexane	298	1.6×10^4
$Bu_3Sn\cdot + Me_3CBr$	cyclohexane	298	3.2×10^8
$Bu_3Sn\cdot + n\text{-}C_5H_{11}Cl$	cyclohexane	298	8×10^2
$Bu_3Sn\cdot + n\text{-}C_5H_{11}Br$	cyclohexane	298	1.8×10^7
$t\text{-}BuO\cdot + Cl_3C\text{-}H$	benzene	300	4.5×10^5
$t\text{-}BuO\cdot + PhCH_2\text{-}H$	benzene	295	8×10^4 (per H)

a: (*continued*)

t-BuO· 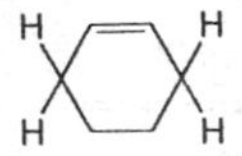 +	benzene	295	1.4×10^{6} (per H)
t-BuO· 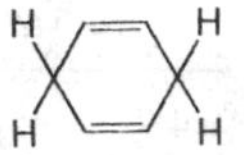 +	benzene	295	1.3×10^{7} (per H)
t-BuO· + PhO-H	benzene	295	3×10^{8}
t-BuOO· + Me_3C-H	isobutane	373	0.6
t-BuOO· + 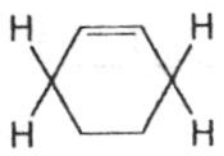	cyclohexene	303	0.1 (per H)
t-BuOO· + 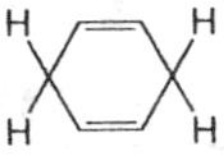	C_6H_5Cl	273	0.5 (per H)

b: Addition reactions

Reacting species	Solvent	Temperature (K)	Rate constant (M^{-1} sec^{-1})
CH_3· + CH_2=$CHCH_3$	H_2O	298	5×10^{3}
CH_3· + CH_2=$C(CH_3)_2$	H_2O	298	4×10^{4}
CH_3· + CH_2=CHCH=CH_2	H_2O	298	1.25×10^{6}
n-alkyl + t-BuN=O	benzene	313	9×10^{6}
n-alkyl 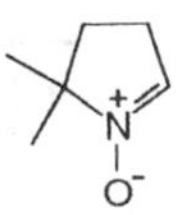 +	benzene	313	2.6×10^{6}
C_6H_5· + C_6H_6	Cl_2FCF_2Cl	298	3.5×10^{5}
C_6H_5· + CH_2=CH(n-alkyl)	Cl_2FCF_2Cl	298	ca. 3×10^{6}
C_6H_5· + C_6H_5CH=CH_2		300	ca. 10^{8}
n-alkyl + O_2		300	ca. 5×10^{9}

c: Unimolecular reactions

Reaction	Solvent	Temperature (K)	Rate constant (sec^{-1})
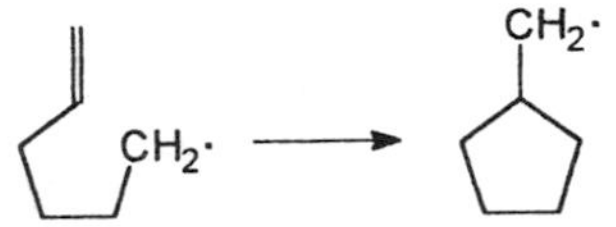 $CH_2\cdot$ → $CH_2\cdot$	benzene	338	1.1×10^5
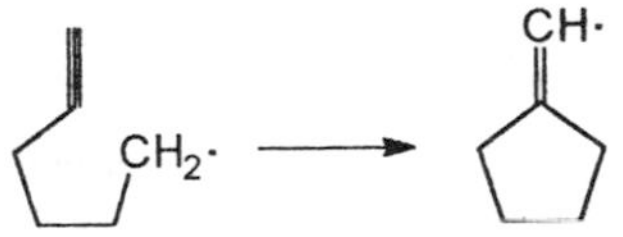 $CH_2\cdot$ → $CH\cdot$	benzene	338	1.5×10^5
 $CH_2\cdot$ → $CH_2\cdot$	C_6F_6	303	2.1×10^8
$PhC(Me_2)CH_2\cdot \rightarrow PhCH_2CMe_2\cdot$	benzene	298	7×10^2
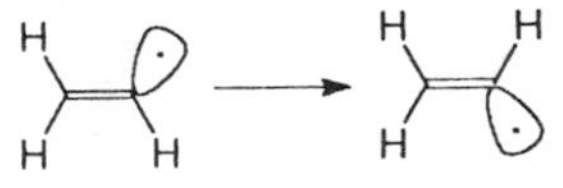 H H H → H H H	C_2H_4	140	$>10^7$
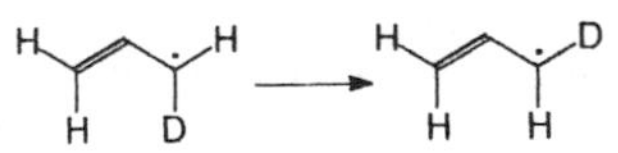 H H H D → H D H H		300	1.3×10^2
t-BuCO· → t-Bu· + CO	methyl-cyclopentane	298	1.2×10^5
$t\text{-}BuCO_2\cdot \rightarrow t\text{-}Bu\cdot + CO_2$			$>10^9$
t-BuOCO· → t-Bu· + CO_2	HCO_2Bu-t	298	3.3×1
 H ·O Me Me Me → HO Me Me Me	cyclohexane	273	4×10^6

Footnotes to tables:

[a] The termination reaction involving two peroxyl radicals is substantially below the diffusion limit.

[b] The product is

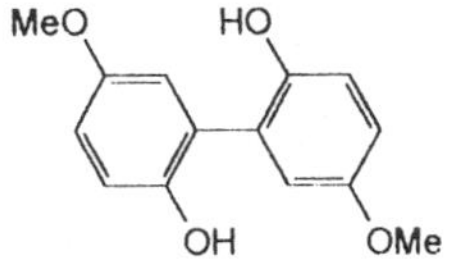

MeO HO OH OMe

[c] Note that the rate of reaction between a highly stabilised nitroxide and a reactive radical is, like those of most of the reactions between two reactive radicals, close to the diffusion-controlled limit. (See Section 9.8, "Stable-radical effects".)

[d] Note that *per molecule* these rates correspond to diffusion control.

[e] Note the solvent effect of benzene compared with the previous entry. (See discussion in Chapter 6, p. 79.)

[f] The very high reactivity of the weak selenol Se–H bond has resulted in the use of benzeneselenol as a scavenger for alkyl radicals. It is worth pointing out that the pK_a of this compound is ca. 4, so that the predominant species in some reaction media may be $PhSe^-$.

Selected additional reading and bibliography

Excellent general information on production, structure, and physical and chemical properties: *Free Radicals* (2 volumes), J.K. Kochi, Ed., Wiley, New York, 1972. A recent basic text pays particular attention to many of the physicochemical aspects of radical properties: *An Introduction to Free Radicals*, by J.E. Leffler, Wiley, New York, 1993.

The subject is regularly reviewed in the annual current awareness series, "Organic Reaction Mechanisms", published by Wiley.

Chapter 2

Rearrangement reactions (including intramolecular addition and atom transfer): A.L.J. Beckwith and K.U. Ingold in *Rearrangements in Ground and Excited States* (Vol. 1), P. de Mayo, Ed., Academic Press, New York, 1984.

S_H2 reactions at multivalent atoms: *Free-Radical Substitution Reactions* by K.U. Ingold and B.P. Roberts, Wiley, New York, 1971.

Chapter 3

Peroxide chemistry: *The Chemistry of Peroxides*, S. Patai, Ed., Wiley, Chichester, 1983 (see also Supplement E2, pub. 1993); *Organic Peroxides*, W. Ando, Ed., Wiley, Chichester, 1992.

Azo-compounds: see *The Chemistry of Hydrazo, Azo, and Azoxy Compounds*, S. Patai, Ed., Wiley, Chichester, 1975.

For a short recent discussion of cage effects which pays particular reference to organometallic systems, see T.W. Koenig, B.P.Hay, and R.G. Finke, *Polyhedron*, **7**, 1499 (1988).

Chapter 4

Selectivity in radical reactions: A.L.J. Beckwith, *Chem. Soc. Reviews*, **22**, 143 (1993).

Steric and polar effects in radical addition: B. Giese, *Angew. Chem. Int. Edn. Engl.*, **22**, 753 (1983). See also the introductory chapter to Giese's book (*vide infra* – Chapter 6).

Stereochemistry of intramolecular radical addition: T.V. RajanBabu, *Accounts Chem. Res.*, **24**, 139 (1991).

Effective molarity in intramolecular radical reactions: C. Berti *et al.*, *Angew. Chem. Int. Edn. Engl.*, **29**, 653 (1990). A more general discussion of this concept in the context of non-radical reactions is given by A.J. Kirby, *Adv. Phys. Org. Chem.*, **17**, 183 (1980); see also L. Mandolini, *ibid.*, **22**, 1 (1986).

Isoselective relationship: B. Giese, *Accounts Chem. Res.*, **17**, 138 (1984).

Chapter 5

There are several excellent texts on electron spin resonance (theory, instrumentation and applications), e.g. *Theory and Applications of Electron Spin Resonance* by W. Gordy, (Techniques of Chemistry series Vol. XV), Wiley, New York, 1980; *Principles of Electron Spin Resonance* by N.M. Atherton, Ellis Horwood, Chichester, 1993; *ESR* by C.D. Poole, Wiley, New York, 2nd edition, 1983; *Modern Pulsed and Continuous-Wave Electron Spin Resonance*, L.Kevan and M.K. Bowman, Eds., Wiley, New York, 1990.

Developments in electron spin resonance and in its use to investigate radical reactions in chemistry and biology have been extensively surveyed in the series of Chemical Society specialist periodical reports entitled "Electron Spin Resonance" published by the Royal Society of Chemistry, London.

Rate constant measurements using ESR: H. Fischer and H. Paul, *Accounts Chem. Res.*, **20**, 200 (1987).

Spin trapping: M.J. Perkins, *Adv. Phys. Org. Chem.*, **17**, 1 (1980).

CIDNP; theory and applications: R. Kaptein, *Advances in Free Radical Chemistry*, **5**, 319 (1975).

Chemical consequences of nuclear spin effects: U.E. Steiner and T. Ulrich, *Chem. Revs.*, **89**, 51 (1989); see also N.J. Turro, *Proc. Nat. Acad. Sci.*, **80**, 609 (1983) and K.A. McLauchlan, *Chem. Soc. Reviews*, **22**, 325 (1993).

Clock reactions: D. Griller and K.U. Ingold, *Accounts Chem. Res.*, **13**, 317 (1980).

Strain in bridgehead radicals: J.S. Lomas, *Accounts Chem. Res.*, **21**, 73 (1988); see also J.C. Walton, *Chem. Soc. Reviews*, **21**, 105 (1992).

Chapter 6

General: *Free Radical Chain Reactions in Organic Synthesis* by D. Crich and W.B. Motherwell, Academic Press, London, 1991. This text contains many useful hints regarding the practice of these reactions as well as their nature and variety.

Solvent effects in chlorination of alkenes: K.U. Ingold, J. Lusztyk, and K.D. Rayner, *Accounts Chem. Res.*, **23**, 219 (1990).

Discussion of some functional group interconversions: D.H.R. Barton, *Aldrichimica Acta*, **23**, 3 (1990); see also *Half a Century of Free Radical Chemistry* by D.H.R. Barton and S.I. Parekh, Cambridge University Press, Cambridge, 1993, and D. Crich and L. Quintero, *Chem. Reviews*, **89**, 1413 (1989).

Formation of new carbon-carbon bonds using radical reactions: *Radicals in Organic Synthesis: Formation of Carbon-Carbon Bonds*, by B. Giese, Pergamon Press, Oxford, 1986.

Radicals from alkylmercurials: G.A. Russell, *Accounts Chem. Res.*, **22**, 1 (1989).

Synthetic applications of inorganic oxidising agents [including NiO_2 and Mn(III)] (see also Chapter 9), "Organic Synthesis by Oxidation with Metal Compounds", W.J. Mijs and C.R.H.I. De Jonge, Eds., Plenum Press, New York, 1986.

Hexenyl radical cyclisations in synthesis: D.P. Curran *Synthesis*, **1988**, 417.

Ring-expansion methodology and annulation reactions involving radicals: P. Dowd and W. Zhang, *Chem. Reviews*, **93**, 2091 (1993).

Radical reactions in natural product synthesis: C.P. Jasperse, D.P. Curran, and T.L. Fervig, *Chem. Reviews*, **91**, 1237 (1991).

Chapter 7

Stereochemical control in radical reactions: N.A. Porter, B. Giese, and D.P.Curran, *Accounts Chem. Res.*, **24**, 296 (1991); see also B. Giese, *Angew. Chem. Int. Edn. Engl.*, **28**, 969 (1989).

Enantioselectivity in hydrogen-atom transfer: P.L.H. Mok, B.P. Roberts, and P.T. McKetty, *J. Chem. Soc. Perkin II*, **1993**, 665.

See also review by T.V. RajanBabu mentioned above (Chapter 4).

Chapter 8

Radical ions: useful background may be found in *Radical Ions*, E.T. Kaiser and L. Kevan, Eds., Wiley, New York, 1968.

Radical cations in pericyclic reactions: N. Bauld, *Accounts Chem. Res.*, **20**, 371 (1987), and *Tetrahedron*, **45**, 5307 (1989); see also H.D. Roth, *Accounts Chem. Res.*, **20**, 343 (1987).

Electron-transfer reactions in organic synthesis: N. Kornblum, *Aldrichimica Acta*, **23**, 71 (1990).

$S_{RN}1$ in aromatic substitution: R.A. Rossi, *Accounts Chem. Res.*, **15**, 164 (1982); see also J.F. Bunnett, *ibid.*, **11**, 413 (1978); **25**, 2 (1992).

Electron transfer in chemistry (Nobel Prize address) by R.A. Marcus, *Angew. Chem. Int. Edn. Engl.*, **32**, 1111 (1993).

Chapter 9

Captodative radicals: H.-G. Viehe, *Accounts Chem. Res.*, **18**, 148 (1985).

Autoxidation of unsaturated lipids: N.A. Porter, *Accounts Chem. Res.*, **19**, 262 (1986); see also C. von Sonntag and H.P. Schuchmann, *Angew. Chem. Int. Edn. Engl.*, **30**, 1229 (1991).

Stable-radical effects: a detailed kinetic analysis is given by H. Fischer, *J. Amer Chem. Soc.*, **108**, 3925 (1986); an interesting inorganic example with a lucid discussion is presented by B.E. Daikh and R.G. Finke, *J. Amer. Chem. Soc.*, **114**, 2938 (1992). However, there seems to be no general review of this important yet little-appreciated phenomenon.

"Substituent effects in radical chemistry": workshop proceedings of that title; H.G. Viehe, Z. Janousek, and R. Merényi, Eds., pub. Reidel, Dordrecht, 1986.

Thermochemistry of radicals: see D.Griller, J.M. Kanabus-Kaminska, and A. Maccoll, *J. Mol. Struct. (Theochem.)*, **163**, 125 (1988), and D. Gutman, *Accounts Chem. Res.*, **23**, 375 (1990).

Chapter 10

Free Radicals in Biology and Medicine: the second edition of an excellent book with this title appeared in 1991, by B. Halliwell and J.M.C. Gutteridge, Oxford University Press, Oxford. In addition, two specialist periodicals are devoted to this field which publish *inter alia* relevant review articles: *Free Radical Biology and Medicine*, and *Free Radical Research Communications*.

DNA and Free Radicals, B. Halliwell and O. Aruoma, Eds., Ellis Horwood, Chichester, 1993.

Recent examples of enzyme studies include the following:

Oxygen rebound in cytochrome P450; V.W. Bowry and K.U. Ingold, *J. Amer. Chem. Soc.*, **113**, 5699 (1991).

Coenzyme B_{12} dependent rearrangements; R.G. Finke, in *Molecular Mechanisms in Bioorganic Processes*, C. Bleasdale and B.T. Golding, Eds., Royal Society of Chemistry, London, 1990, p. 244.

Ribonucleotide reductase; J. Stubbe *et al.*, *ibid.*, p. 305.

Isopenicillin N synthase; R.M. Adlington, *ibid.*, p. 1; J.E. Baldwin and C. Schofield, *The Chemistry of β-lactams*, M.I. Page, Ed., Chapman and Hall, New York, 1992, p.1.

Pyruvate formate lyase; J. Knappe *et al.*, *Biochem. Soc. Trans.*, **21**, 731 (1993); this is one of a series of review articles based on a symposium on radical-dependent enzyme mechanisms.

Photosynthesis: Excellent discussions of the reaction centres of photosynthetic bacteria are examined in the Nobel lectures of J. Diesenhofer and H. Michel, and R. Huber, *Angew. Chem. Int. Edn. Engl.*, **28**, 829, 848 (1989).

The physiological role of NO: A.R. Butler and D.L.H. Williams, *Chem. Soc. Reviews*, **22**, 213 (1993); P.L. Feldman, O.W. Griffith, and D.J. Stuehr, *Chemical and Engineering News*, **1993**, December 20th, p. 26.

Enediyne antibiotics: K.C. Nicolaou and W.-M. Dai, *Angew. Chem. Int. Edn. Engl.*, **30**, 1387 (1991); K.C. Nicolaou and A.L. Smith, *Accounts Chem. Res.*, **25**, 497 (1992).

Problems

Problem-solving is an essential aid to comprehension. The problems set out in the following pages have been selected to enable the reader to test his or her understanding of the principles outlined in this text. However, in a few instances, the opportunity has been taken to introduce some principles and applications of radical chemistry not discussed elsewhere in the book. For these, as for most of the problems, references have been provided so that solutions can be compared with published ones;[1] in some cases hints as to the accepted solution are included.

[1]It should be remembered that published interpretations of experimental data are not invariably correct! Should you consider that a different rationalisation of any results published in the literature (or elsewhere in this book!) is preferable to the one given, consider what experiments you might perform in order to exclude one of the alternatives.

1. t-Butoxyl radicals from t-butyl peroxyoxalate react with alkyl benzyl ethers, $PhCH_2OR$, to give products which include the dimers $[PhCHOR]_2$, (**1**), and benzaldehyde, (**2**). Explain how these are formed, and account for the variation in ratio (**1**)/(**2**) with R which is indicated below. (Hint: see footnote 2 on p. 60.)

R	Me	Et	Me_2CH	Me_3C	$PhCH_2$	1-Adamantyl
(1)/(2)	4.5	2.2	1.6	0.4	0	2.5

2. Write down the contributing resonance structures for the radical anions of nitrobenzene and *p*-benzoquinone. Note the structural similarity between the former species and phenyl nitroxide (PhNHO·).

3. The following transformations proceed by radical-chain mechanisms. In each case suggest plausible propagating steps.

a: $BrCCl_3$, initiator (AIBN); products: Br, CCl_3

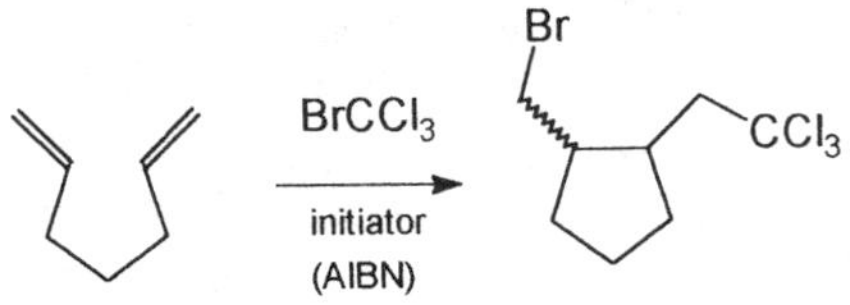

b: initiator (AIBN), CCl_4; O, N, S, O; Cl + N, S, CCl_3

c: $BrCCl_3$, initiator (AIBN); Br, CCl_3 + Br, CCl_3; minor product

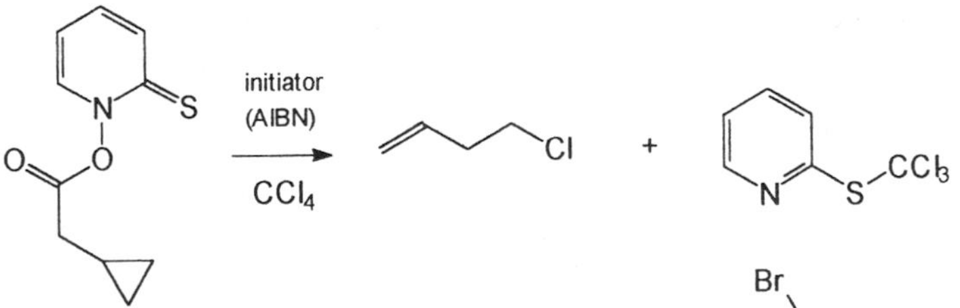

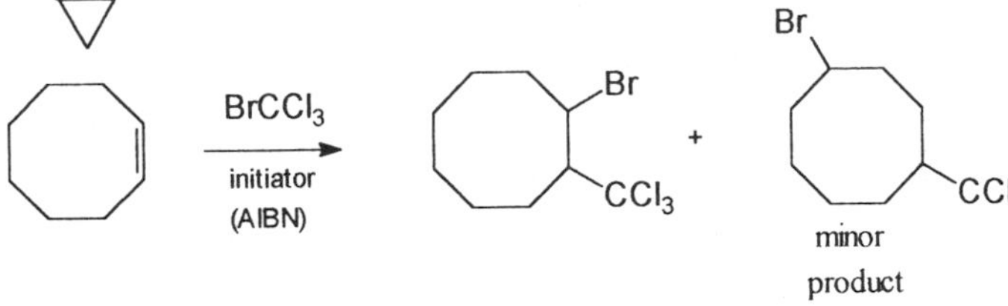

J. Org. Chem., **32**, 529 (1967).

d: O; PhSSPh (0.3 equiv), initiator (AIBN), *hv*; O

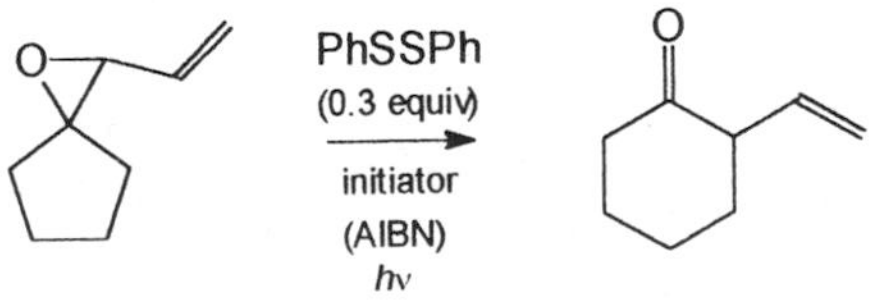

Tetrahedron Letters, **1991**, 6575.

e: I, MeN, SO_2, N; Bu_3SnH, initiator (AIBN); MeNH, N

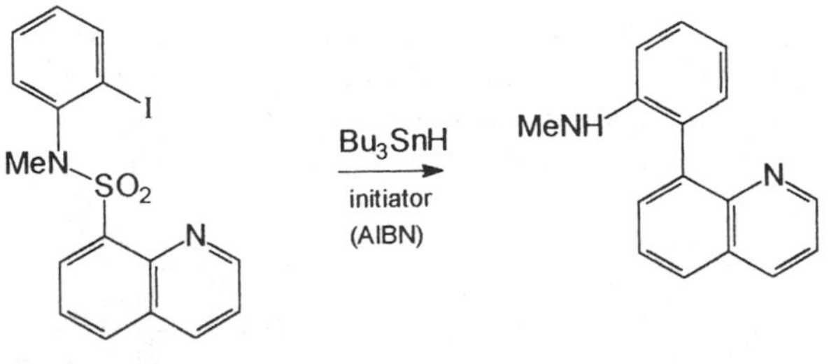

3 (*continued*)

f: OEt, O, I, t-BuMe2SiO; Bu_3Sn–CH=CH–C(O)–C_5H_{11}, initiator → OEt, O, C_5H_{11}, O, t-BuMe2SiO

g: Bu_3SnH (cat. quantity), t-BuNC (30 equivalents) → OEt, O, CN, t-BuMe2SiO

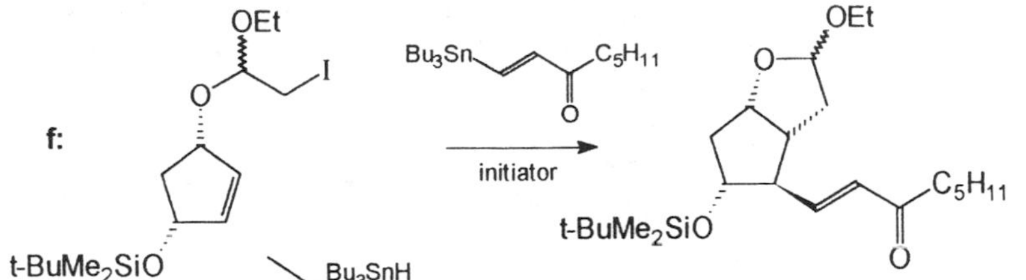

J. Org. Chem., **52**, 2958 (1987).

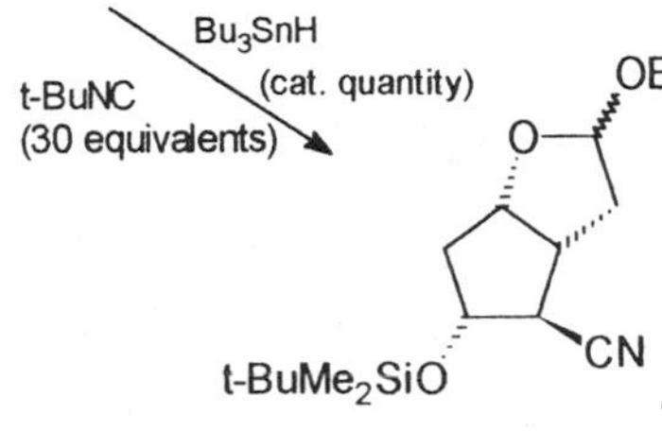

J. Amer. Chem. Soc., **108**, 303 (1986).

h: $h\nu$/40°C, O_2 → OOH; $h\nu$ → O

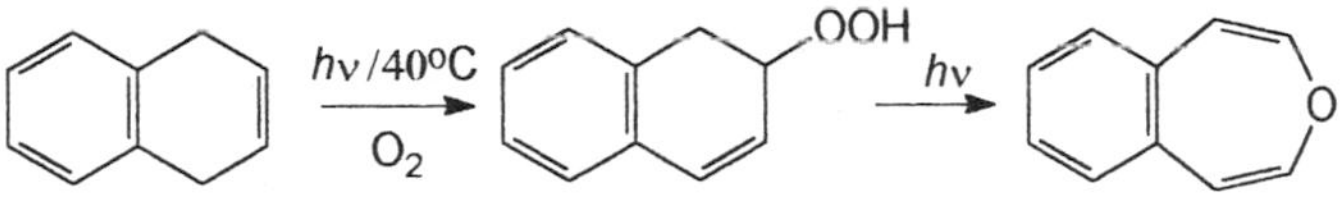

J. Amer. Chem. Soc., **94**, 4048 (1972).

i: O, Me, I, $SnBu_3$; Bu_3SnH (cat), initiator → O, Me

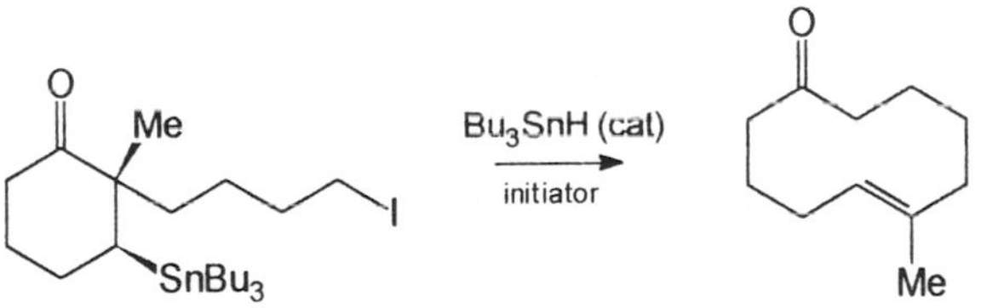

Tetrahedron, **1991**, 6795.

j: S, N, O, O, N, Bu; $h\nu$, BF_3OEt_2 1 equiv. → Bu, N, S, N

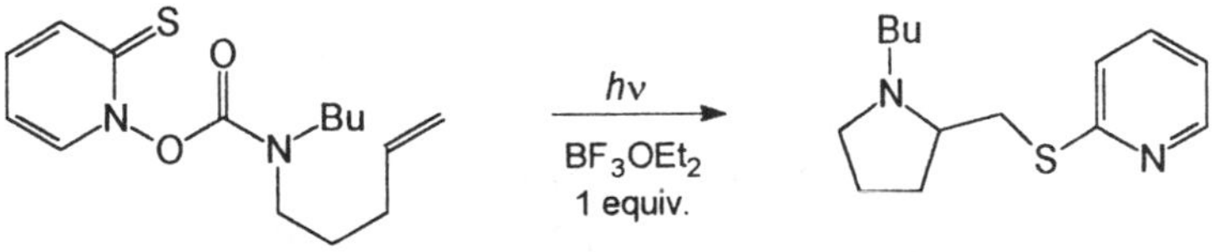

(N.B. pyrrolidine formation is negligible in the absence of Lewis acid)

Tetrahedron Letters, **1991**, 6493.

In part **c**, replacement of the $BrCCl_3$ by CCl_4 gives similar adducts ($C_9H_{14}Cl_4$), except that the 1,4-isomer predominates. Explain this difference.

In part **i**, when the butyl side chain is *cis* to the tributyltin substituent the *Z*-isomer of the cyclodecenone is obtained. Discuss the observed stereospecificity of these reactions.

4. On being heated in boiling chlorobenzene, compound **(1)** gives **(2)** as well as biphenyl and a mixture of all three isomeric monochlorobiphenyls. Write out a reaction scheme which explains the formation of each of these products.

Ph Ph H H Ph Ph → Ph Ph H Ph

(1) **(2)**

Tetrahedron Letters, **1971**, 2379.

5. Explain the formation of each of the products shown in the following reactions.

a: Ph—C(Me)(Me)—CH_2CH_2CHO $\xrightarrow[100^o]{t\text{-BuO-OBu-t}}$

Ph—C(Me)(Me)—CH_2CH_3 + [tetralone] + PhC(O)—$CH_2CH_2CH(CH_3)_2$

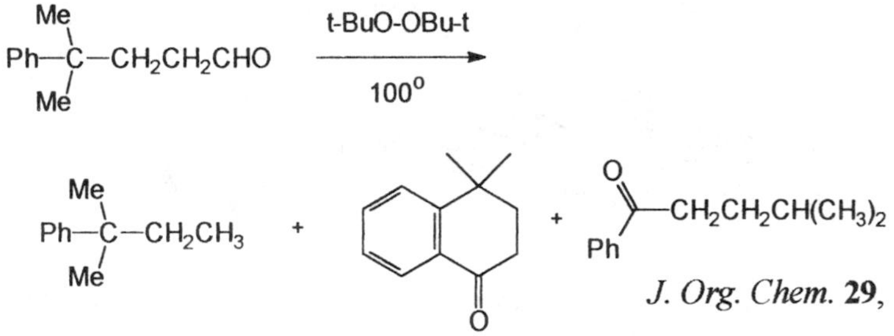

J. Org. Chem. **29**, 1663 (1964).

b: OOH $\xrightarrow[\text{HOAc/ H-donor}]{FeSO_4}$ O + O

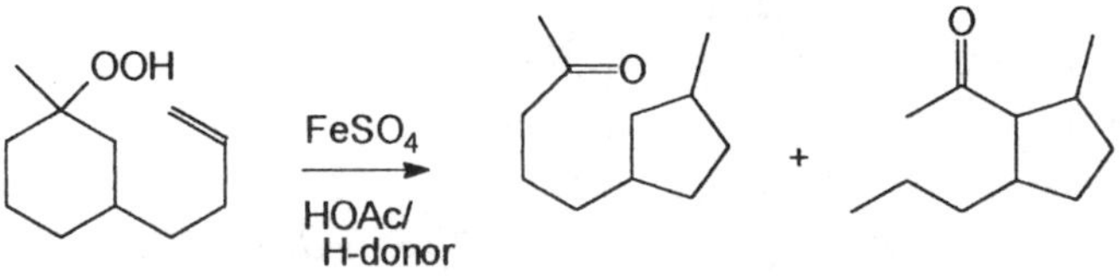

Tetrahedron Letters, **1986**, 5981.

c: Br $\xrightarrow{Bu_3SnH \text{ (high conc)}}$

$\xrightarrow{Bu_3SnH \text{ (low conc)}}$

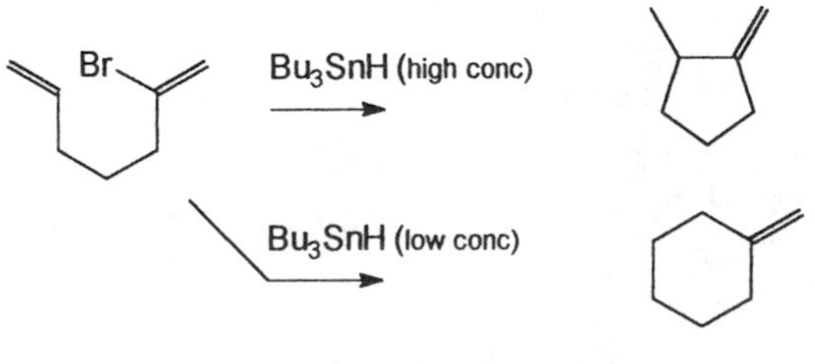

Tetrahedron Letters, **1986**, 4525; 4529.

d: Me—C≡C—$CH_2CH_2CH_2\cdot$ + :C=N—C_6H_4—F → Me, F, N + Me, N, F

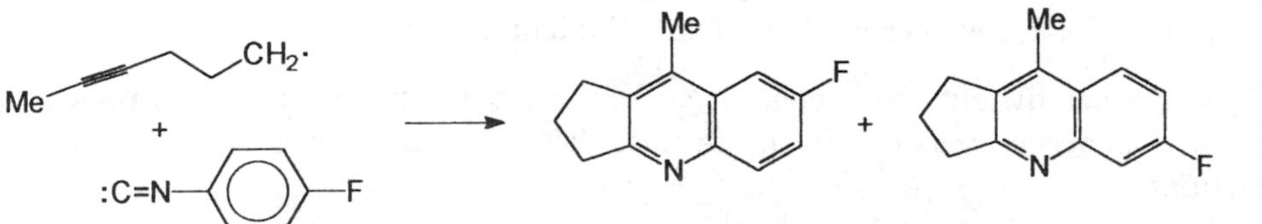

J. Amer. Chem. Soc., **113**, 2127 (1991).

5 (*continued*)

e: Br $\xrightarrow[\text{AIBN}]{Bu_3SnH}$ O

J. Amer. Chem. Soc., **113**, 939 (1991).

f: O O $\xrightarrow[\text{AIBN}]{Bu_3SnH}$ O $SnBu_3$ O H

J. Amer. Chem. Soc., **112**, 902 (1990).

g: S O N N Me Me $SiMe_3$ $\xrightarrow[\text{AIBN}]{Bu_3SnH}$ Me Me $SiMe_3$

J. Chem. Soc. Chem. Comm., **1988**, 1380.

h: O Me N_3 $\xrightarrow[\text{AIBN}]{Bu_3SnH}$ (followed by removal of Bu_3Sn substituent) O H N Me

J. Amer. Chem. Soc., **115**, 3328 (1993).

i: As well as phenol, reaction of hydroxyl radicals with benzene gives biphenyl. (Hint: see Section **9.7**.)

J. Chem. Soc., **1964**, 4857.

6. Even when carried out using high-dilution techniques to minimise the reaction of the uncyclised intermediate radical with Bu_3SnH, the reduction of the iodoacetamide (**1**) gives more of the simple reduction product (**2**) than of the desired hydroindanone (**3**). What factors, other than competition between reduction of the cyclised intermediate and its cyclisation, may be important here?

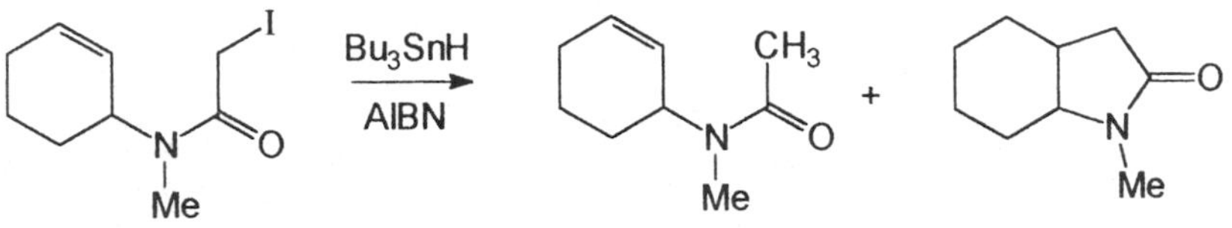

Bu3SnH
AIBN
CH3
+
N O Me
(**1**) (**2**) (**3**)

When, instead, a solution of (**1**) in benzene containing 0.55 equivalents of $Bu_3SnSnBu_3$ and three equivalents of ethyl iodide is exposed to ultraviolet irradiation, a 70% isolated yield of (**4**) may be obtained. Explain.

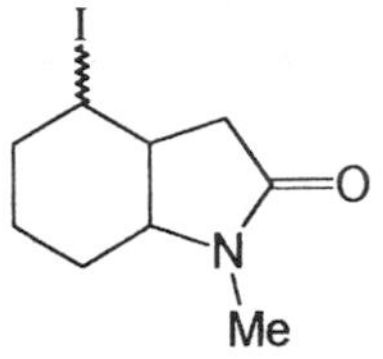

(**4**)

(Two important new principles are illustrated in these reactions.)

J. Amer. Chem. Soc., **110**, 7536 (1988).

7. When a moderately concentrated solution of a well-known initiator is heated in t-butylbenzene (a high-boiling, rather unreactive solvent) in the microwave cavity of an ESR spectrometer, at such temperature that the half-life of the initiator is very short, it is possible to detect a 21-line spectrum which is simulated below. What is the initiator?

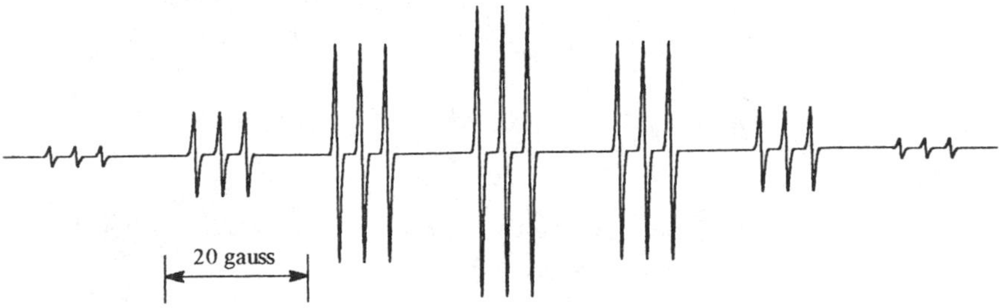

20 gauss

8. Suggest a mechanism for the formation of tricyclic nitroxide (**1**) by nickel peroxide (NiO_2) oxidation of phenol in the presence of nitrosobenzene.

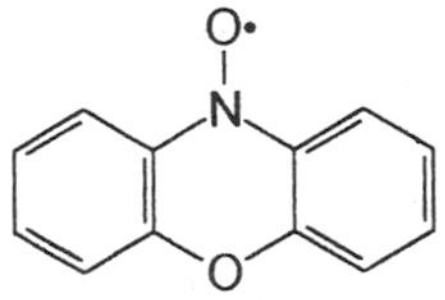

(1)

Chemistry Letters, **1972**, 115.

9. Possible radical reactions are proposed from time to time in biosynthetic pathways. Examples include alkaloids derived from phenol oxidation (Chapter 6) and the prostaglandins (Chapter 10). A plant metabolite which is an insect "antifeedant" has structure (**1**) in which the usual steroid 5-membered D-ring has been expanded and aromatised. The biosynthesis of (**1**) was thought likely to involve the co-metabolite (**2**) as a precursor, which might initially be oxidised to the diene (**3**) and then, after oxidation at the angular methyl group, might rearrange by a radical mechanism. To test this hypothesis, the model compound (**4**) was photolysed using visible light. One product proved to be (**5**). Write a mechanism which accounts for the formation of (**5**).

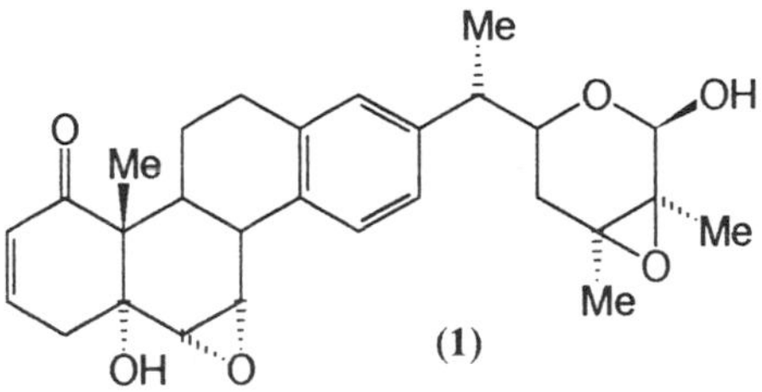

(1)

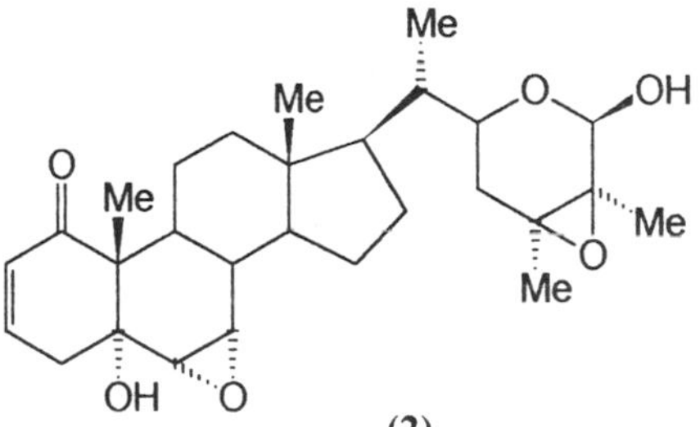

(2)

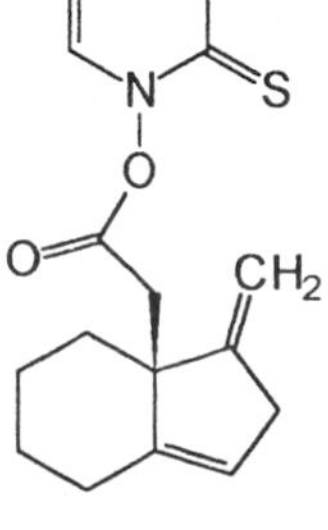

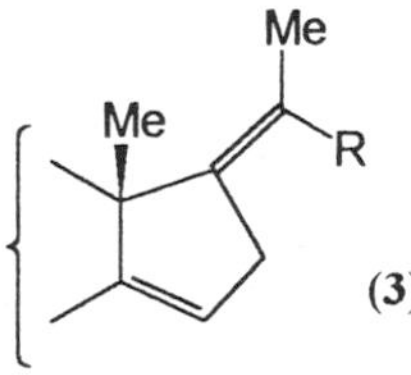

(3)

J. Chem. Soc. Chem. Comm. **1992**, 1754.

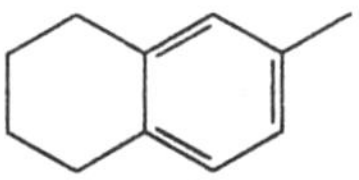

(4)

(5)

10. The possibility of generating the succinimidyl radical from *N*-bromosuccinimide (NBS) or other precursors has attracted much interest over the years, and many approaches to its detection have been investigated, including spin trapping. Exposure of NBS and 2-methyl-2-nitrosopropane (MNP) in benzene to visible light gives the nine-line spectrum shown in Fig. **1**. This has reasonably been interpreted as being due to (**1**). Suggest how the radical responsible for this spectrum may arise. Do the cautionary remarks given in Chapter 5 regarding spin trapping suggest the possibility of any alternative mechanistic interpretations?

When the peroxyester (**2**) was heated with MNP in oxygen-free chlorobenzene, and the solution cooled, the spectrum shown in Fig. **2** was observed. A clue to the origin of this was obtained when the reaction mixture was heated in the spectrometer cavity. Under these conditions the 27-line spectrum was superimposed on spectra of several other species, one of which was unmistakably that of the methyl spin adduct (Me-MNP·), i.e. t-BuN(Me)O·. What species is probably responsible for the 27-line spectrum, and what experiments might you carry out to test whether your interpretation is correct? (Hint: consider the discussion of nitroxide-radical reactions outlined in Section 5.4.)

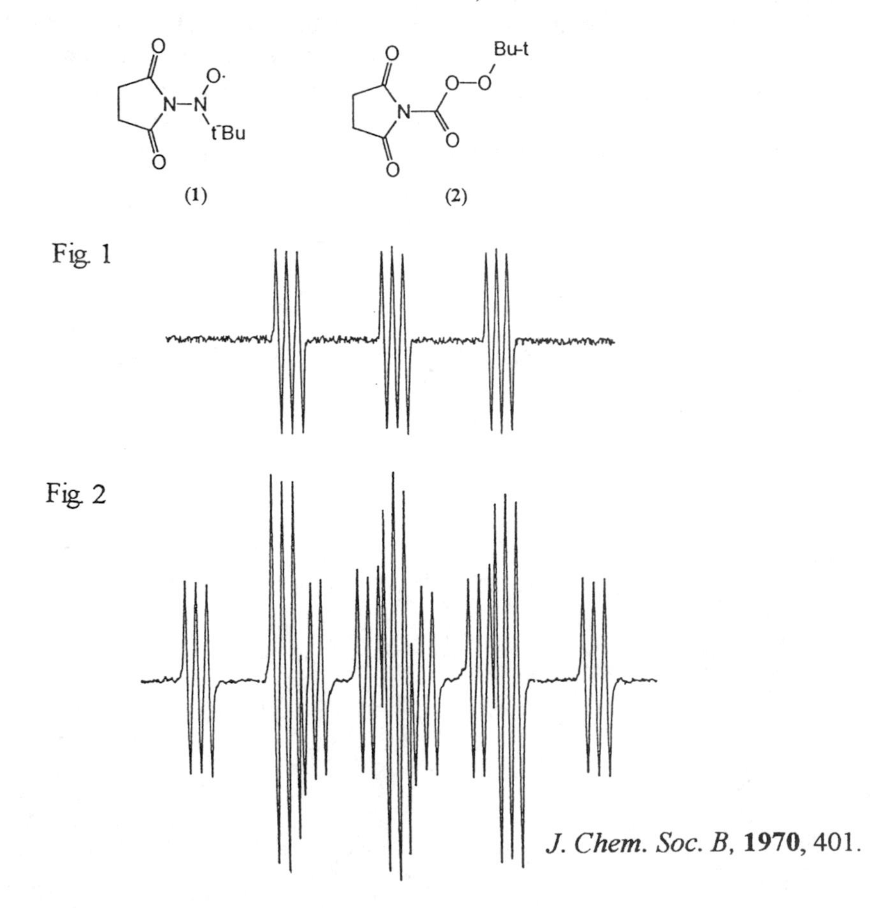

J. Chem. Soc. B, **1970**, 401.

11. At 60°, the first order decomposition of t-butyl perbenzoate is slower by a factor of 2.5×10^5 than that of its *o*-thiophenyl derivative (**1**). Suggest an explanation.

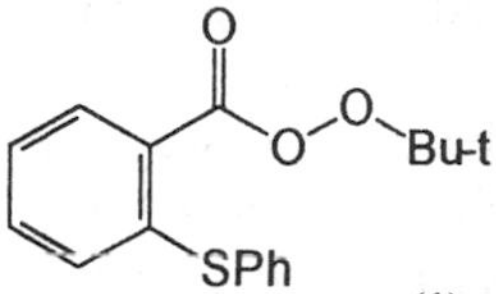

O
O O Bu-t
SPh
(**1**)

J. Amer. Chem. Soc., **82**, 1561 (1960).

12. In synthetic applications of the hexenyl radical cyclisation and related processes, an important factor in determining the efficiency of the reaction is the competition between ring closure of the hexenyl radical and its bimolecular reaction with reagent. For example, in the reduction of 6-bromo-1-hexene by tri-n-butyltin hydride, if the concentration of tin hydride is too high a significant yield of 1-hexene accompanies the methylcyclopentane.

Using the rate constants for hexenyl cyclisation and for reaction of primary alkyl radicals with tri-n-butyltin hydride given in the Appendix, construct a plot of yield ratio (methylcyclopentane/hexene) as a function of tin hydride concentration. Assume that the tin hydride is in large excess, so that its concentration may be regarded as constant. What concentration of tin hydride is necessary for equal amounts of these two products to be formed?

13. Addition of a THF solution of the allyl ether, (**1**), and acetophenone to a solution of SmI_2 in THF/HMPA gave, after acidification, (**2**), in poor yield (<20%). However, if the acetophenone was added 5 minutes after (**1**), the yield of (**2**) was 89%.

In a second reaction, addition of the ether was followed by addition of CuI and then 2-cyclohexenone. In this case the conjugate addition product, (**3**), was obtained.

How can these synthetically significant results be accounted for?

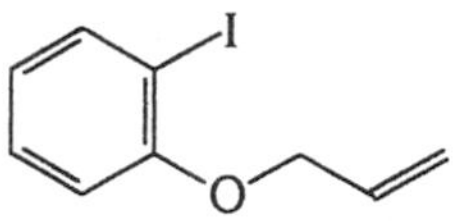

I
O
(**1**)

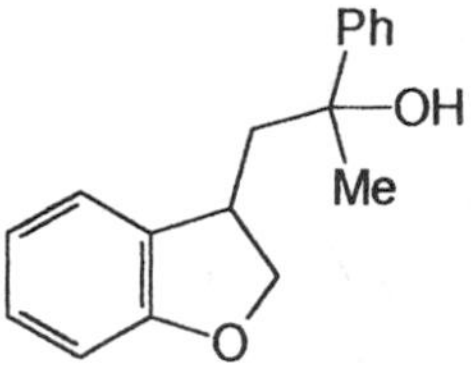

Ph
OH
Me
O
(**2**)

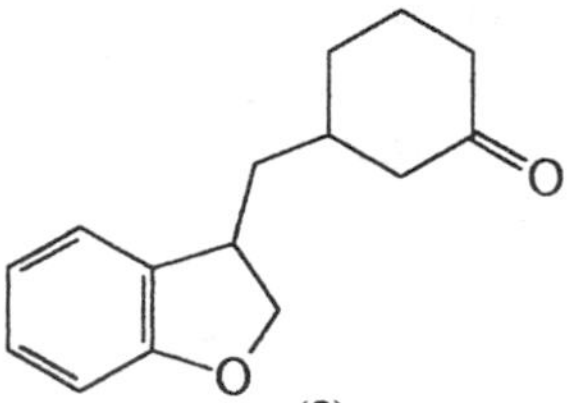

O
O
(**3**)

J. Amer. Chem. Soc., **114**, 6050 (1992); *J. Org. Chem.*, **57**, 1740 (1992).

14. Unlike the P450 enzymes (Chapter 10), bacterial methane monooxygenases (which catalyse the oxidation of methane to methanol) have non-haem iron centres. Investigation of one of these using the radical-clock method has yielded interesting results, some of which are outlined below, although their interpretation is still somewhat ambiguous.

In contrast to the result for the P450 system described in Chapter 10, oxidation of racemic *trans*-dimethylcyclopropane gives the cyclopropylcarbinol (**1**), with no evidence for the formation of the ring-opening product (**2**). Apparently, therefore, if a radical mechanism (e.g. "oxygen rebound") operates, the radical lifetime must be even shorter than with the P450 enzyme.

Oxidation of racemic *trans*-1-methyl-2-phenylcyclopropane (**3a**) also gives unrearranged product (**4**), accompanied by (aromatic) ring-hydroxylated products (i.e. phenols), but no ring-opened material. The rate of ring opening of the cyclopropylcarbinyl radical (**6**) has been shown to be three orders of magnitude different from that of opening of (**5**). Would you expect it to be faster or slower, and what then appears to be the significance of this result? Can you think of any alternative explanation that may account for the result with (**3a**)?

A kinetic isotope effect (k_H/k_D) of about 5 was observed for oxidation of (**3b**), but the components of a mixture of (**3a**) and (**3c**) were oxidised at essentially identical rates. Furthermore, it was found that the same ratio of phenols to unrearranged carbinol was obtained from (**3a**) as was obtained from (**3c**). Consider possible explanations for these data. (Note: racemic hydrocarbon was used in these experiments; it was not established whether or not the carbinol (**4**), produced in this reaction, was also racemic).

Me, CH_2OH (**1**) OH (**2**)

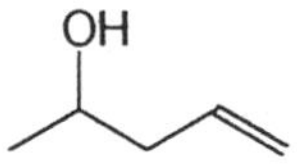

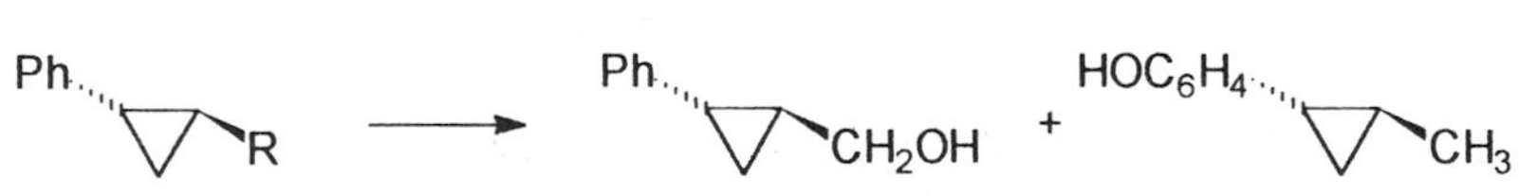

Ph, R → Ph, CH_2OH + HOC_6H_4, CH_3

(**3**) **a**: R=CH_3
b: R=CH_2D
c: R=CD_3

(**4**)

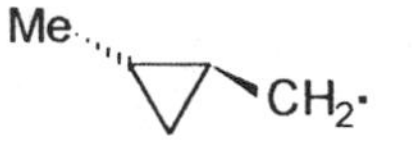

Me, $CH_2\cdot$ (**5**) Ph, $CH_2\cdot$ (**6**)

J. Amer. Chem. Soc., **115**, 939 (1993).

15. It was pointed out in Chapter 6 (p. 87) that the efficiency of tandem radical cyclisation to form *bicyclic* product from *acyclic* precursor is limited by the stereochemistry of the first cyclisation step. If this first step were to form a cycloalkyl radical rather than a cycloalkylmethyl radical, the stereochemical problem would be eliminated. One way of achieving this is illustrated in the example given below, which also incorporates a potentially valuable new departure in azide chemistry. Write a mechanism which accommodates this transformation.

Interestingly, simple alkyl radicals do not react intermolecularly with the azide function of alkyl azides. Why, then, does the intramolecular reaction succeed? (See Section 4.7.)

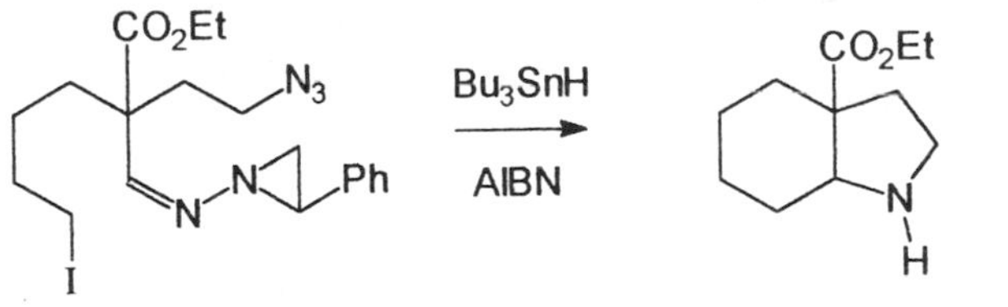

CO_2Et N_3 N N Ph I

$\xrightarrow[\text{AIBN}]{Bu_3SnH}$

CO_2Et N H

(>70%)

J. Amer. Chem. Soc., **116**, 5521 (1994).

Subject index